T0321946

# Material
# Geometry

*Groupoids in Continuum Mechanics*

# Material Geometry
## Groupoids in Continuum Mechanics

**Manuel de León**
*Consejo Superior de Investigaciones Científicas, Spain*

**Marcelo Epstein**
*University of Calgary, Canada*

**Víctor Jiménez**
*Universidad de Alcalá (UAH), Spain*

NEW JERSEY · LONDON · SINGAPORE · BEIJING · SHANGHAI · HONG KONG · TAIPEI · CHENNAI · TOKYO

*Published by*

World Scientific Publishing Co. Pte. Ltd.

5 Toh Tuck Link, Singapore 596224

*USA office:* 27 Warren Street, Suite 401-402, Hackensack, NJ 07601

*UK office:* 57 Shelton Street, Covent Garden, London WC2H 9HE

**Library of Congress Cataloging-in-Publication Data**
Names: León, Manuel de, 1953–    author. | Epstein, M. (Marcelo), author. |
    Jiménez, Víctor (Víctor Manuel), author.
Title: Material geometry : groupoids in continuum mechanics /
    Manuel de León, Consejo Superior de Investigaciones Científicas, Spain,
    Marcelo Epstein, University of Calgary, Canada,
    Víctor Jiménez, Universidad de Alcalá (UAH), Spain.
Description: Hackensack, New Jersey : World Scientific, [2021] |
    Includes bibliographical references and index.
Identifiers: LCCN 2021003618 | ISBN 9789811232541 (hardcover) |
    ISBN 9789811232558 (ebook for institutions) | ISBN 9789811232565 (ebook for individuals)
Subjects: LCSH: Continuum mechanics. | Geometry, Differential. | Groupoids. | Algebroids.
Classification: LCC QA808.2 .L466 2021 | DDC 531.01/51228--dc23
LC record available at https://lccn.loc.gov/2021003618

**British Library Cataloguing-in-Publication Data**
A catalogue record for this book is available from the British Library.

For any available supplementary material, please visit
https://www.worldscientific.com/worldscibooks/10.1142/12168#t=suppl

Desk Editors: Vishnu Mohan/Lai Fun Kwong

Typeset by Stallion Press
Email: enquiries@stallionpress.com

Printed in Singapore

*This book is dedicated to the extraordinary generation of founders of modern continuum mechanics, who were driven by a scientific goal and an aesthetic ideal that will not be easy to replicate.*

*"Ut proinde his paucis consideratis tota haec materia redacta sit ad puram Geometriam, quod in physicis & mechanicis unice desideratur."*

*"So that, having considered these few things, this whole matter may be reduced to pure geometry, which in physical and mechanical subjects is especially desired."*

# Preface

In *continuum physics* the physical properties of an elastic body are characterized for all the constitutive relations. These measure the mechanical response produced at each particle by a deformation in a local neighborhood of the particle. Differential geometry provides a rigorous mathematical framework not only to present the constitutive properties but also to discover and prove results. For applications, it is usual that the bodies are assumed uniform and homogeneous in the sense of that the body is made of a unique material and there is a configuration in such a way that the mechanical response is the same at all the points.

In this book, we use the Noll's approach to present a mathematical framework based on groupoids, algebroids and distributions to deal with non-uniform and inhomogeneous simple bodies.

For any simple body, a unique groupoid, called *material groupoid*, may be naturally associated. The uniformity of the body coincides with the transitivity of the groupoid. If the material groupoid turns out to be a Lie groupoid, the associated Lie algebroid, called *material algebroid*, is available. Then, the homogeneity is characterized by the integrability of both (material groupoid and material algebroid).

However, the property of being Lie groupoid is not guaranteed for any elastic material. In fact, smooth uniformity corresponds to that imposition of differentiability on the material groupoid. Smooth distributions are now introduced to deal with this case. In fact, two smooth distributions, called *material distributions*, may be canonically defined generalizing the notion of Lie algebroid. Thus, it is proved that we can cover the simple body by a material foliation whose leaves are (smoothly) uniform. These new tools are also used to present a "measure" of uniformity and a homogeneity for non-uniform bodies.

The construction of the material distribution is generalized to a much more abstract framework in which the case of an arbitrary subgroupoid of a Lie groupoid is considered. This more general theory is used to prove a number of new results. Among others, we prove that any subset $N$ of a manifold $M$ may be endowed with a maximal foliation in such a way that $N$ is a submanifold if and only if $N$ is a leaf of this foliation.

# About the Authors

**Manuel de León** was born in Requejo (Zamora, Spain) in 1953. He received his master degree from the University of Santiago de Compostela (Galicia) and then the PhD degree in the same university in 1978. His work has been mainly devoted to the study of differential geometry and its applications to mechanics and mathematical physics (over 300 papers and 4 monographs). He has been the advisor for 12 PhD students, and is currently a member of several Editorial Boards and Scientific Committees and Editor-in-Chief of *Journal of Geometric Mechanics (AIMS)*. He was Research Professor in the Consejo Superior de Investigaciones Científicas (CSIC), and Chief of the Department of Mathematics (CSIC), 2000–2007. He was also the Vice-Director of the Institute of Mathematics and Fundamental Physics (CSIC), 1992–1998. He has been the founder and director of the Instituto de Ciencias Matemáticas (ICMAT), a Center of Excellence in Spain. He is also fellow of the Real Academia de Ciencias Exactas, Físicas y Naturales, and the Real Academia Canaria de Ciencias and Real Academia Galega de Ciencias. He founded the series of International Fall Workshops on Geometry and Physics. He was refounder and vice-president of the Real Sociedad Matemática Española (RSME), and director of La Gaceta de la RSME. He was also coordinator of the Spanish Committee for the World Mathematical Year 2000.

He was the founder and first president of the Spanish Committee for Mathematics CEMAT), 2004–2007, chair and president of the International Congress of Mathematicians ICM2006-Madrid, member of the Executive Committee of the International Mathematical Union, IMU (2007–2014) and member of the Executive Board of the International Council for Science, ICSU, now ISC (2014–2018). He has been a member of several national and international committees for assessment of science (PESC Committee, European Science Foundation; Comisión Asesora de Evaluación y Prospectiva, MICINN; Comisión de Area de Ciencias Físicas y Tecnologías Físicas, CSIC).

**Marcelo Epstein** was born in Buenos Aires, Argentina and graduated as a Civil Engineer from the University of Buenos Aires. He obtained a doctorate from the Technion, Israel Institute of Technology. Since 1976, he has been a professor of mechanical engineering at the University of Calgary, Canada. He has published extensively in the field of continuum mechanics and in the area of biomechanics, where theories of growth and remodeling provide creative opportunities for the application of the theory of groupoids to the description of evolving materials. Some of his previous books include: *Material Inhomogeneities and Their Evolution* (Springer, 2007; co-authored with M. Elzanowski), *The Geometrical Language of Continuum Mechanics* (Cambridge University Press, 2010), *The Elements of Continuum Biomechanics* (Wiley, 2012), and *The Latin of Science* (Bolchazy-Carducci, 2019). He is a fellow of the American Academy of Mechanics, a recipient of the CANCAM medal (2009), the Tullio Levi-Civita award in Differential Geometry and Mechanics (2014), and the APEGA Frank Spragins Technical Summit Award (2020).

**Víctor Jiménez,** born in 1988 in Santa Cruz de Tenerife. He did his undergraduate studies and his Master in Mathematics at Universidad de La Laguna (ULL). He became member of ICMAT in December 2015, supported by a four year FPI grant, where he completed his PhD studies in November 2019. His dissertation deals with the application of abstract mathematical tools like groupoids or distribution on Continuum Mechanics with Professors Manuel de León and Marcelo Epstein. Since then, he has published 9 papers not only on this topic, but on nonholonomic mechanics and connections on Lie algebroids. In September 2020, he got a position in the Univesity of Alcalá, where he is developing his research as an Assistant Professor.

# Contents

# Chapter 1

# Introduction

From its origins, the theory of elasticity has been a rich and exciting branch of mathematical research. This subject was initially created by J. Bernoulli, A. L. Cauchy and L. Euler and there is a long list of important mathematicians who made substantial contributions to this branch: Beltrami, Birkhoff, Hadamard, Lipschitz, among others.

Over the years, however, the importance of elasticity as a branch of mathematics was decreasing. Although there were still exceptions, in general mathematicians lost interest in elasticity (see Wang and Truesdell, 1973).

In 1954 the thesis of W. Noll entitled '*On the Continuity of the Solid and Fluid States*' (Noll, 1954) gave a new impulse to the theory. Here, W. Noll started to use the concept of *material points*. This subject, which could be called *new rational elasticity*, is the physical basis of this book.

In particular, we will be interested in the interaction between Continuum Mechanics and Differential Geometry. As may be found in the modern books *Introduction to Rational Elasticity*, due to C. C. Wang and C. Truesdell, and *Mathematical Foundations of Elasticity*, due to J. E. Marsden and T. J. R. Hughes, this relation has resulted in a very rich theory full of interesting results. It is remarkable that this relationship has even an older history. In fact, theories of elastic beams and shells had already needed the use of results of differential geometry of curves and surfaces. The work of the Cosserat brothers anticipated certain aspects of modern differential geometry by the addition of microstructure to the material body.

The clearest link between these two areas is provided by the fact that a *continuum* is physically modelized as a three-dimensional connected manifold $\mathcal{B}$, the *material body*, which can be embedded in $\mathbb{R}^3$, i.e., $\mathcal{B}$ has a global chart. The physical space is identified with $\mathbb{R}^3$. Observe that a material body is, by definition, an abstract topological space. Then, as a matter of applications, we need to depict the body as a part of the physical world. To deal with this problem, we define the *configurations*. A configuration $\phi$ is an embedding from $\mathcal{B}$ into $\mathbb{R}^3$. We usually fix a reference configuration $\phi_0$. Then, a *deformation* is a change of configurations, namely the composition $\phi \circ \phi_0^{-1}$.

In a more abstract formulation (Segev, 1986), many of the limitations of the theory, such as the metric structure of space, can be removed without affecting the physical part of the theory. In fact, in Segev (1986) a rigorous theory arises from imitating the geometric approach of Classical Mechanics by defining the configuration space as the space of embeddings of the body into the physical space. This configuration space has a (non-unique) structure of an infinite-dimensional differentiable manifold (Hirsch, 1994; Kriegl and Michor, 1997).

In this book, we will work on a different facet of the interaction between Continuum Mechanics and Differential Geometry. The theory describing the elastic fields of certain kind of isolated defects, which are now called *dislocations* and *disclinations*, was originally developed by Vito Volterra in 1907 (Volterra, 1907). However, it was not until 1955 that a rigorous theory of continuous distributions of defects was conceived by K. Kondo (Kondo, 1955), D. A. Bilby (B. A. Bilby, 1960), E. Kröner (Kröner, 1960, 1968), J.D. Eshelby (Eshelby, 1951) and others (see also the books Lardner (1974) and Nabarro (1987)). This structurally based theory is motivated by heuristic considerations, mostly studying limiting process starting from a defective crystalline structure.

There is another distinct approach proposed by W. Noll (Noll, 1967). Although both approaches use similar geometric structures, the conceptual status of the theory, however, is quite different. The Noll's approach is based on the existence of *constitutive laws* encoding all the information about the material response of the body. This permits us to compare material points via the so-called *material isomorphisms* (notion which will be discussed below). Although some of the results achieved by this school of thought are the same as those of its predecessors, we can also find important differences. The fundamental role of the *material symmetry groups* in Noll's theory is one of them.

The notion of material isomorphism is the main idea of the Noll's theory. Denote by $W$ the mechanical response of the body $\mathcal{B}$, such as the *Cauchy stress*, or the elastic energy per unit mass. Let us assume that $\mathcal{B}$ is a *simple elastic body*, i.e., the constitutive laws are completely characterized at a point by itself and the gradient of the deformation at that point. Then, the mechanical response is represented as a differentiable map $W : \mathcal{B} \times Gl(3, \mathbb{R}) \to V$ (where $Gl(3, \mathbb{R})$ is the general linear group of regular $3 \times 3$-matrices and $V$ is a finite-dimensional vector space). Notice that the form of $W$ depends on the choice of a particular reference configuration $\phi_0$ to express the gradient of the deformations.

Let us identify the body $\mathcal{B}$ with some particular reference configuration, and let $X, Y \in \mathcal{B}$ be two particles. $X$ and $Y$ are said to be *materially isomorphic* if there exists a linear isomorphism $P_{XY} : T_X\mathcal{B} \to T_Y\mathcal{B}$ such that

$$W(Y, F) = W(X, FP_{XY}), \tag{1.1}$$

for all deformation gradients $F$. In this case $P_{XY}$ is called a *material isomorphism between $X$ and $Y$*. If $X = Y$, then $P_{XX}$ is called a *material symmetry at $X$*. The set $G(X)$ of all the material symmetries at $X$ is, in fact, a group called the *material symmetry group*.

The idea of materially isomorphic body points provides mathematical rigor to the physical property of being made of the same material. A body $\mathcal{B}$ is *materially uniform* if all its points are made of the same material or, equivalently, if all its points are materially isomorphic. So, we may claim that two body points $X$ and $Y$ are materially isomorphic if and only if their constitutive functions are the same except for a local change of reference configuration.

Material isomorphisms $P_{XY}$ are not, in general, unique. In fact, the set of material isomorphisms $G(X, Y)$ from $X$ to $Y$ may only satisfy one of these two conditions:

- $G(X, Y) = \emptyset$.
- $G(X, Y)$ is in a one-to-one correspondence with the material symmetry group $G(X)$ at $X$. In fact,

$$P_{XY} \cdot G(X) = G(X, Y),$$

for any material isomorphism $P_{XY} \in G(X, Y)$.

Taking into account these facts and assuming that the body is *smoothly uniform*, Noll introduced the notion of *material parallelisms* and their associated curvature-free *material connections*. This idea was further extended by C. C. Wang (Wang, 1967) and F. Bloom (Bloom, 1979). We present here another approach to these notions (see also Jiménez *et al.*, 2019).

The non-vanishing of the torsion of the (not necessarily unique) material connections plays an important role in Noll's formulation. In the case of discrete symmetry groups, the material connection is unique and, then, we have a *"canonical"* tensor (the torsion of the connection) measuring the presence of material defects. In Noll's terminology, the identical vanishing of the torsion of the connection is tantamount to the notion of *local homogeneity* of the body, which is physically interpreted as the absence of defects.

The frame bundle of the body and, more particularly, its *G*-structures provide a new formulation of these ideas. In Elżanowski *et al.* (1990), M. Elżanowski, M. Epstein and J. Śniatycki associate to any smoothly uniform body a family of conjugated *G*-structures, *material G-structures*, in such a way that their flatness characterizes the homogeneity of the material. However, this approach has certain limitations. In particular, the material *G*-structures are not canonically defined and they are not amenable to generalization for the description of defects in for non-uniform materials.

The staging of groupoids solves these two problems. Since the notion of groupoid is not so familiar for many readers, let us give here a brief description in order to explain the essence of this relevant generalization of the concept of group. The *groupoid* is somewhat of a latecomer to the mathematical repertoire, its earliest appearance and naming being traceable to an article by Heinrich Brandt (1886–1954) published in 1927 in the prestigious *Mathematische Annalen*. It is clear that Brandt had an original idea, and that he was eager to communicate it. The very title of Brandt's paper hints at his basic idea and choice of terminology: *On a generalization of the concept of group*. Brandt's definition consists of introducing in a set, just as in the case of a group, a binary internal operation (the *product* or *composition*), but (and this is the essential difference between a group and a groupoid) this operation is *not necessarily defined for all ordered pairs* of elements. In other words, given two elements $a, b$, the composition $c = ab$ may, or may not, exist.

This operation, however, is not arbitrary, but must satisfy a number of algebraic properties (such as associativity and existence of inverses and some special elements which act as right and left units of sorts).

Brandt's definition, formulated in the spirit of an algebraic generalization of the concept of group, can be replaced by an equivalent one which cleverly describes the elements of a groupoid as *arrows* between points of an underlying set. This definition, with its clear geometric and physical overtones, turns out to be a creative tool for the discovery of applications, such as the one implied in the main topic of this book. Let us, therefore, consider a *total set* $\Gamma$, whose elements are the arrows, as our point of departure. Each arrow $g \in \Gamma$ has a tail (or *source*) and a head (or *target*), which will be considered as elements of another set, the set of *objects* $M$.

Pictorially, we may imagine the total set $\Gamma$ as a cloud of arrows hovering in place above a meadow $M$. These two sets, $\Gamma$ and $M$, are linked by means of two *projection* maps, $\alpha : \Gamma \to M$ and $\beta : \Gamma \to M$, called, respectively, the *source map* and the *target map*. As indicated by their names, $\alpha(g)$ and $\beta(g)$ give us, respectively, the tail $x$ and the tip $y$ of the arrow $g$. This is shown schematically in Fig. 1.1, but it is important to take these representation with a grain of salt. Arrows, tips, and tails, are just figures of speech. Thus, an arrow does not quite have intermediate points, and the tip and the tail of an arrow in $\Gamma$ belong to the set $M$. All that the arrow imagery conveys is that there is a certain "relation" between $x$ and $y$. An elegant and suggestive notation for a groupoid is $\Gamma \rightrightarrows M$.

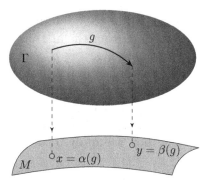

Fig. 1.1.   A groupoid $\Gamma \rightrightarrows M$ as a cloud of arrows hovering over a meadow $M$.

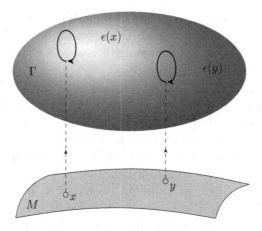

Fig. 1.2.    The section of identities $\epsilon : M \to \Gamma$ and the identities.

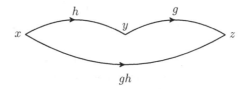

Fig. 1.3.    Tip-to-tail composition.

The projection maps, $\alpha$ and $\beta$, map $\Gamma$ onto $M$. The *section of identities* $\epsilon$, on the other hand, goes in the opposite direction, from $M$ into $\Gamma$. It assigns to each object $x$ a very special arrow, called the *identity* at $x$, such that $\alpha(\epsilon(x)) = \beta(\epsilon(x)) = x$. Thus, these identities are represented as loop-shaped arrows, as shown in Fig. 1.2.

Given two points, $x$ and $y$, in the underlying set $M$ there may, or may not, be an arrow connecting them. But if there is an arrow $h$ connecting $x$ to $y$, and another arrow $g$ connecting $y$ to $z$, then surely there is the arrow $gh$ connecting $x$ to $z$. Intuitively, the two original arrows are composed in the natural "end to tail" fashion. The notation $gh$ (rather than $hg$) is more suggestive, in the sense that we first apply $h$, as it were, to the point of departure $x$, thus "arriving" at $y$, and only then we apply $g$ to get to $z$, as suggested in Fig. 1.3.

For every arrow $g$, there is necessarily an inverse arrow going in the opposite direction. This feature of a groupoid is clearly reminiscent of

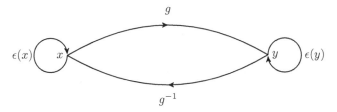

Fig. 1.4.   Inverse and identities.

the similar idea in the case of a group. It corresponds to the notion of reflexivity of a relation. Finally, since the identities are loop-shaped arrows, they can be composed to the left with all arriving arrows, and to the right with all departing ones, leaving those arrows unchanged. These properties, represented in Fig. 1.4, can be expressed as

$$g\,\epsilon(\alpha(g)) = \epsilon(\beta(g))\,g = g \quad gg^{-1} = \epsilon(\beta(g)) \quad g^{-1}g = \epsilon(\alpha(g)). \quad (1.2)$$

Not all loop-shaped arrows are units. To understand this point more clearly, we may ask the question: is a group a particular case of a groupoid and, if so, how? If the underlying set $M$ happens to be a singleton (a set with just one element), then surely all arrows must be mutually composable! Moreover, there is just a single identity arrow. Thus, a group is a groupoid with a singleton as its base set, as depicted in Fig. 1.5.

In general, for each object $x$ in $M$, we can consider the subset $\Gamma_x^x$ consisting of all loop-shaped arrows that start and end at $x$. More formally,

$$\Gamma_x^x = \{g \in \Gamma \mid \alpha(g) = \beta(g) = x\}. \quad (1.3)$$

This set is never empty, since it contains at least the identity at $x$. Moreover, all the arrows in $\Gamma_x^x$ are mutually composable. It follows from a trivial application of the definition of a groupoid that each $\Gamma_x^x$ is a group, called the *isotropy group* at $x$.

There is now one important technical point that can be made without using much additional mathematical apparatus. Suppose that two different points, $x$ and $y$, of $M$ are connected by an arrow $g$ (such that $\alpha(g) = x$ and $\beta(g) = y$). It should be intuitively expected that the local symmetries, embodied in the respective isotropy groups $\Gamma_x^x$ and $\Gamma_y^y$, are in some sense 'the same'. Let $g$ belong to $\Gamma_x^x$. We can think of it as a loop-shaped arrow at $x$. But then, according to the

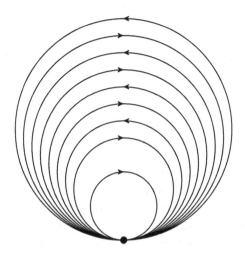

Fig. 1.5.   A group as a groupoid over a singleton.

tip-to-tail paradigm, the arrow $h' = ghg^{-1}$ is a loop-shaped arrow at $y$.[1] Since we can reverse the argument by working in the opposite direction, we can convince ourselves that this operation, when applied to every element of either one of the isotropy groups, is a bijection. Simply put, there is a one-to-one invertible correspondence between the local symmetries at $x$ and those at $y$, and this correspondence is of the form $h' = ghg^{-1}$ for some fixed $g$. In the parlance of group theory, we may say that if two points of $M$ are connected by at least one arrow, the corresponding isotropy groups are *conjugate*. In general, a groupoid is said to be *transitive*, if all the points are connected by one arrow. Consequently, if a groupoid is transitive, all of its isotropy groups are mutually conjugate. Choosing any of these groups $G$ as a model, we can say that a transitive groupoid is endowed with a *structure group* $G$. This is the closest that a groupoid can get to a group without actually being one.

The narrative of this book begins with the recognition that the collection of all material isomorphisms $P_{XY}$ for all pairs of body points $X, Y$ of a material body $\mathcal{B}$ (uniform or not) is a groupoid, called *material groupoid*, which is a subgroupoid of the

---

[1]A simple diagram, similar to that used in Fig. 1.4, can be drawn to reinforce the argument.

1-jets groupoid $\Pi^1(\mathcal{B}, \mathcal{B})$ on $\mathcal{B}$ (Epstein and de León, 1998, 2016). The material groupoid of $\mathcal{B}$ will be denoted by $\Omega(\mathcal{B})$. Therefore, uniformity and homogeneity will be studied by using the properties of the material groupoid.

A first consequence of the structure of groupoid of $\Omega(\mathcal{B})$ is the following: $\mathcal{B}$ is uniform if and only if $\Omega(\mathcal{B})$ is transitive. If the material groupoid $\Omega(\mathcal{B})$ is a Lie subgroupoid of $\Pi^1(\mathcal{B}, \mathcal{B})$, the *associated Lie algebroid* $A\Omega(\mathcal{B})$ is available. Therefore, a series of results are presented characterizing the homogeneity by the properties of the Lie algebroid (Jiménez *et al.*, 2019).

We will prove $\Omega(\mathcal{B})$ is a subgroupoid of $\Pi^1(\mathcal{B}, \mathcal{B})$. A natural question now arises: Is always $\Omega(\mathcal{B})$ a Lie subgroupoid of $\Pi^1(\mathcal{B}, \mathcal{B})$? The answer is negative. We will prove that $\mathcal{B}$ is smoothly uniform if and only if $\Omega(\mathcal{B})$ is a transitive Lie subgroupoid of $\Pi^1(\mathcal{B}, \mathcal{B})$. Explicit examples of non-uniform bodies in which the material groupoid is not a Lie subgroupoid of $\Pi^1(\mathcal{B}, \mathcal{B})$ are presented in this book.

Thus, we should face the case in which $\Omega(\mathcal{B})$ is simply an algebraic subgroupoid of $\Pi^1(\mathcal{B}, \mathcal{B})$. Even in that case, we may generalize the construction of the associated Lie algebroid and construct a smooth distribution of $\Pi^1(\mathcal{B}, \mathcal{B})$ called *material distribution* $A\Omega^T(\mathcal{B})$ (see Jiménez *et al.*, 2017 or Jiménez *et al.*, 2018). $A\Omega^T(\mathcal{B})$ is generated by the (local) left-invariant vector fields on $\Pi^1(\mathcal{B}, \mathcal{B})$ which are in the kernel of $TW$. Due to the groupoid structure, we can still associate two new objects to $A\Omega^T(\mathcal{B})$, denoted by $A\Omega(\mathcal{B})$ and $A\Omega^\sharp(\mathcal{B})$, as defined by the following diagram:

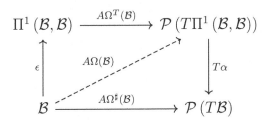

Here $\mathcal{P}(E)$ defines the power set of $E$, $\epsilon(X)$ is the identity map of $T_X\mathcal{B}$ and $\alpha : \Pi^1(\mathcal{B}, \mathcal{B}) \to \mathcal{B}$ denotes the source map of the groupoid.

By construction, the distributions $A\Omega^T(\mathcal{B})$ and $A\Omega^\sharp(\mathcal{B})$ are integrable (in the sense of Stefan (1974) and Sussmann (1973)) and they provide two foliations, $\overline{\mathcal{F}}$ on $\Pi^1(\mathcal{B}, \mathcal{B})$ and $\mathcal{F}$ on $\mathcal{B}$, such that $\Omega(\mathcal{B})$ is union of leaves of $\overline{\mathcal{F}}$. As a consequence, we have that $\mathcal{B}$ can be covered by a foliation of some kind of smoothly uniform "sub-bodies", called *material submanifolds*. The material distribution also offers a tool able to provide a general classification of smoothly non-uniform bodies and the possibility to distinguish various degrees of uniformity. In addition, homogeneity may be generalized in such a way that any simple body can be checked for homogeneity.

Next, we may consider a more general situation. We study the problem from a purely mathematical framework, since we are convinced that this analysis should be relevant not only for its applications to Continuum Mechanics, but also for the general theory of groupoids.

So, let $\overline{\Gamma} \subseteq \Gamma$ be a subgroupoid of a Lie groupoid $\Gamma \rightrightarrows M$; notice that we are not assuming, in principle, any differentiable structure on $\overline{\Gamma}$. Even in that case, we can construct a generalized distribution $A\overline{\Gamma}^T$ over $\Gamma$ generated by the (local) left-invariant vector fields on $\Gamma$ whose flow at the identities is totally contained in $\overline{\Gamma}$. This distribution $A\overline{\Gamma}^T$ will be called the *characteristic distribution of* $\overline{\Gamma}$. Again, due to the groupoid structure, we can still associate two new objects to $A\overline{\Gamma}^T$, denoted by $A\overline{\Gamma}$ and $A\overline{\Gamma}^\sharp$, called the *base-characteristic distributions of* $\overline{\Gamma}$, analogously to the above diagram.

The relevant fact is that both distributions, $A\overline{\Gamma}^T$ and $A\overline{\Gamma}^\sharp$, are integrable and they provide two foliations, $\overline{\mathcal{F}}$ on $\Gamma$ and $\mathcal{F}$ on $M$. Studying the properties of these foliations we obtain the following two main results, Theorem 4.4 and Theorem 4.2, which may be summarized in the following statement:

**Theorem** *Let* $\Gamma \rightrightarrows M$ *be a Lie groupoid and* $\overline{\Gamma}$ *be a subgroupoid of* $\Gamma$ *(not necessarily a Lie groupoid) over* $M$. *Then, there exists a maximal foliation* $\overline{\mathcal{F}}$ *of* $\Gamma$ *and a maximal foliation* $\mathcal{F}$ *of* $M$ *such that*

(i) *for all* $x \in M$,
$$\alpha\left(\overline{\mathcal{F}}(x)\right) = \mathcal{F}(x);$$

(ii) $\overline{\Gamma}$ *is a union of leaves of* $\overline{\mathcal{F}}$;

(iii) *for all* $x \in M$ *there exists a transitive Lie subgroupoid* $\overline{\Gamma}(\mathcal{F}(x))$ *of* $\Gamma$ *with base* $\mathcal{F}(x)$.

So, although our groupoid $\overline{\Gamma}$ is not a Lie subgroupoid of $\Gamma$, we can still cover it by manifolds (leaves of the foliation $\overline{\mathcal{F}}$) and extract the "transitive" and "differentiable" components (the Lie groupoids $\overline{\Gamma}\left(\mathcal{F}\left(x\right)\right) \rightrightarrows \mathcal{F}\left(x\right)$).

The book may be divided into two parts clearly differentiated. The first part (Part I: Fundamentals) is devoted to present an introduction to the non-elementary but necessary knowledge to understand the rest of the book, and is divided into three chapters, Continuum Mechanics, Groupoids, and Algebroids, and each of these parts is intended to the reader who is not an expert on the subject. The goal of this chapter is to make the book as self-contained as possible to facilitate for the reader the comprehension of the material.

In this way, Chapter 2 is responsible for introducing the reader into the world of Continuum Mechanics. In particular, we will focus on a brief review of the *Constitutive Theory of Materials* for simple elastic bodies.

Groupoids are the topic of Chapter 3. We start the chapter with the analysis of a groupoid induced by a real game, *15-puzzle groupoid*, to help the reader to get an idea of what a groupoid is. Then, we present a rigorous theory of groupoids focusing on the results which we are interested in. We also give several examples of groupoids. A particularly relevant example for this book is the 1-jets groupoid. As a last introductory chapter, Chapter 4 is devoted to Lie algebroids. Similarly to the previous chapter, we start with a simple example of Lie algebroid, the tangent algebroid, to give an idea of what a Lie algebroid is. Next, the category of Lie algebroids is constructed presenting a consistent theory of Lie algebroids with several examples. An explicit construction of Lie's functor from Lie groupoids to Lie algebroids is then studied. A first example is the *1-jets algebroid*. So, we show a Lie algebroid isomorphism from the 1-jets Lie algebroid to the algebroid of derivations which gives us a different way to depict the 1-jets algebroid. As a final part of this chapter, we study the Lie's fundamental theorems for Lie groupoids and Lie algebroids. We begin studying Lie's first fundamental theorem, which shows that any integrable Lie algebroid can be integrated by a Lie groupoid with simply connected $\beta$-fibres. Then, with a similar development, a theorem can be proved about integrability of subalgebroids. On the other hand, we will describe Lie's second fundamental theorem which studies the integration of Lie algebroid morphisms and we will use this to prove some consequences which will be useful in the book. We conclude by giving an example of a non-integrable Lie algebroid to show that Lie's third fundamental theorem is not true for Lie algebroids.

The beginning of the second part serves as an introduction and a motivation to the study. In particular, we present the material groupoid which can be considered the *cornerstone* of this book. We also prove two results characterizing the (smooth) uniformity over the material groupoid which give us an intuition about the path which we have taken.

The second part shows the consequences of using the theory of groupoids in the constitutive theory of materials. This part is divided into two chapters: The first chapter exhibits the results obtained as a consequence of assuming that the material groupoid is a Lie subgroupoid of the 1-jets groupoid. In the second part, we study the material groupoid in a more general sense (without the assumption of regularity imposed in the previous chapter) introducing new tools to deal with this case.

Some of the results presented in Chapter 5 were published in Jiménez *et al.* (2019) for simple materials. Here we calculate the material algebroid using the algebroid of derivations to give more than one representation of this Lie algebroid. We use this to characterize, above all, the homogeneity in different ways.

Chapter 6 is devoted to the general case in which there are not assumptions on the material groupoid (Epstein *et al.*, 2019; Jiménez *et al.*, 2017; Jiménez *et al.*, 2018; Jiménez *et al.*, 2020). We divide this chapter into three sections. The first section deals with the general case of a subgroupoid $\overline{\Gamma}$ (not necessarily a Lie subgroupoid) of a Lie groupoid $\Gamma \rightrightarrows M$. We then construct the so-called characteristic distribution which provides us with "pseudo-differentiable" structure on the subgroupoid $\overline{\Gamma}$ generalizing the structure of Lie groupoid. In particular, we obtain a way to give a "pseudo-differentiable" structure on any subset $N$ of a manifold $M$ generalizing (in some "maximal" way) the structure of smooth manifold. In Section 6.2, we apply all the results of the previous section to simple materials. In this case, characteristic distributions will have a particular shape and will be called material distributions. Thus, as an interesting result, we obtain that any body $\mathcal{B}$ can be covered by a maximal foliation of some kind of "(smoothly) uniform subbodies". We also present a "measure of uniformity", the *graded uniformity*, based on the material distributions. Finally, a notion of homogeneity for non-uniform bodies is introduced and characterized in different ways. As we predicted, the material groupoid will not have to be a Lie subgroupoid of the 1-jets groupoid. Thus, in Section 6.3 we give some examples of this. We study the graded uniformity and the homogeneity on these examples.

Finally, we present some appendices that include the necessary notions for the book to be self-contained. First, in Appendix A we present the concepts of foliations and distributions. Here we prove some classical integrability theorems which are used in the book. Next, Appendix B we give an introduction to covariant derivatives and their relation with distributions. Finally, Appendix C is devoted to *principal bundles*. This appendix will be focused on the so-called *frame bundles*. Here, we also introduce the notion of *connections in principal bundles*.

# Part I
# Fundamentals

## Chapter 2

# Continuum Mechanics: Elastic Simple Bodies

We will start with a very brief sketch of the indispensable background in *Continuum Mechanics*. A more detailed study may be found in the books (Epstein, 2010; Epstein and Elzanowski, 2007; Wang and Truesdell, 1973). Another recommendable reference is Marsden and Hughes (1994).

To move from classical mechanics of finite systems of particles to mechanics of continuum materials arises the problem of finding a proper definition of *body*. In this book, we are interested in the so-called *deformable body*. Such a model defines a *body* as an oriented manifold $\mathcal{B}$ of dimension 3 which can be covered by just one chart. The points of the manifold $\mathcal{B}$ will be called *body points* or *material particles* and will be denoted by using capital letters ($X, Y, Z \in \mathcal{B}$). A *subbody* of $\mathcal{B}$ is an open subset $\mathcal{U}$ of the body.

It is important to note that, by definition, a material body (with its material particles) is just a topological space. So, one could think that the body lives in some *"abstract world"* outside the reality. To manifest the material inside the *"real world"* we will introduce the so-called *configurations*.

**Definition 2.0.1.** A configuration of a body $\mathcal{B}$ is given by an embedding $\phi : \mathcal{B} \to \mathbb{R}^3$. The 1-jet $j^1_{X,\phi(X)}\phi$ at the body point $X \in \mathcal{B}$ (see Appendix C) is called an infinitesimal configuration at $X$.

Thus, a configuration of a material body assigns to any particle $X$ a spot in the space in a smooth way such that two particles cannot be assigned to the same spot. Points on the Euclidean space $\mathbb{R}^3$ will be called *spatial points* and will be denoted by lower case letters $(x, y, z \in \mathbb{R}^3)$.

We usually identify the body with one of its configurations, say $\phi_0$, called *reference configuration*. Coordinates in the reference configuration will be denoted by $X^I$, while any other coordinates will be denoted by $x^i$.

In spite of the choice of the reference configuration, any formulation should turn out to be independent of this choice. The established physical rules should not depend on the representation of the body in the real or physical world.

A statement or property on the material body $\mathcal{B}$ is said to be *configuration indifferent* or *configuration independent* if it does not depend on the chosen reference configuration (*uniformity* in Definition 2.0.5 will be a good example of configuration independent property).

**Definition 2.0.2.** Given any arbitrary configuration $\phi$, the change of configurations $\kappa = \phi \circ \phi_0^{-1}$ is called a *deformation*, and its 1-jet $j^1_{\phi_0(X), \phi(X)} \kappa$ is called an *infinitesimal deformation at* $\phi_0(X)$.

Notice that, by using the *Polar decomposition theorem*, any infinitesimal deformation at a body point may be decomposed as follows,

$$F = RU = VR, \qquad (2.1)$$

where $R$ is an orthogonal matrix and $U$ and $V$ are symmetric positive-definite tensors. $R$ is called the *rotation tensor* and $U$ and $V$ the *right* and *left stretch tensors of* $F$. An important circumstance results when the determinant of $F$ is positive. In that case, the rotation tensor $R$ is a *pure rotation* (no mirror needed). On the other hand, any (real) positive definite matrix may be diagonalized, i.e., it is similar to a diagonal matrix given by its eigenvalues. So, the physical interpretation of Eq. (2.1) is that the effect of an infinitesimal deformation at a point on a vector is to apply three stretches (given by the eigenvalues of $U$) and a rotation given by $R$. Analogously we interpret $F = VR$. Observe that, Eq. (2.1) permits us to separate the strain information ($U$ or $V$) from the rotation information ($R$).

The *right* and *left Cauchy–Green tensors of F* are given by $C = F^T F = U^2$ and $B = FF^T = V^2$. If two infinitesimal deformations have the same right (or left) stretch tensor, one follows from the other by a rotation.

In term of local coordinates, by following notation introduced above, the entries of the matrix $F$ are denoted as follows:

$$F^i_J = \frac{\partial x^i}{\partial X^J} \tag{2.2}$$

where the transformation is the composition of the local coordinates $(x^i)$ with the inverse of the reference configuration $\phi_0 = (X^I)$. Then, the Cauchy–Green tensors have the following expression:

$$C_{IJ} = F^i_I F^i_J$$
$$B^{ij} = F^i_I F^j_I.$$

The Einstein summation convention will be used along the whole book. There is still another important tensor to introduce: the *Lagrangian strain tensor* which is given by the formula:

$$E = \frac{1}{2}(C - I), \tag{2.3}$$

where $I$ is the identity matrix (or the identity tensor in the reference configuration).

To be able to predict the deformation of a body in motion is an important goal in continuum mechanics. It is very intuitive that the internal constitution of the body should play a role. For instance, steel, wood or gel will not be deformed equally when they are subject to the same loading. The mathematical interpretation of this is that the dynamical principles alone should not be enough to determine the motion of a deformable body. In fact, the response of the body to the history of its deformations is supposed to be characterized for one or more *constitutive equations*.

The experiments seem to indicate that the material response is a *local property* in such a way that a material particle is "*affected*" only for what is inside of a small neighborhood of the particle. There also exists a non-local treatment of continuum media initiated by Eringen (1972a,b, 1983, 2002) based on the assumption that the stress on a material particle depends on all the body points of the continuum material.

*Elastic simple bodies* (Wang and Truesdell, 1973) are character-ized under the assumption that the constitutive law depends on a point only on the infinitesimal deformation at the same point. Thus, a *mechanical response* for an elastic simple material $\mathcal{B}$ in a given refer-ence configuration $\phi_0$ is mathematically formalized as a differentiable map $W$ from the set $\mathcal{B} \times Gl(3, \mathbb{R})$, where $Gl(3, \mathbb{R})$ is the general linear group of $3 \times 3$-regular matrices, to a fixed (finite dimensional) vector space $V$. In this book, we will not be interested in the nature of the vector space $V$. However, in general, $V$ will be the space of *stress tensors*. In fact, in continuum mechanics, the contact forces at a par-ticle $X$ in a given configuration $\phi$ are characterized by a symmetric second-order tensor $T_{X,\phi}$ on $\mathbb{R}^3$ called the stress tensor. Thus, $T_{X,\phi}$ is a linear map

$$T_{X,\phi} : \mathbb{R}^3 \to \mathbb{R}^3.$$

From a physical point of view, $T_{X,\phi}$ turns the unit normal of a smooth surface into the stress vector acting on the surface at $\phi(X)$. Then, the mechanical response is given by the following identity:

$$W(X, F) = T_{X,\phi},$$

where $F$ is the 1-jet at $\phi_0(X)$ of $\phi \circ \phi_0^{-1}$.

We should now clarify how the mechanical response changes with the changing of reference configuration. Let $\phi_1$ be another configura-tion and $W_1$ be the mechanical response associated to $\phi_1$. Then, we will impose that for any other (local) configuration $\phi$

$$W(X, F_0) = W_1(X, F_1), \tag{2.4}$$

where $F_i$, $i = 0, 1$, is the associated matrix to the 1-jet at $\phi_i(X)$ of $\phi \circ \phi_i^{-1}$. Hence, obviously Eq. (2.4) implies that

$$W_1(X, F) = W(X, F \cdot C_{01}), \tag{2.5}$$

for all regular matrix $F$, where $C_{01}$ is the associated matrix to the 1-jet at $\phi_0(X)$ of $\phi_1 \circ \phi_0^{-1}$. So, Eq. (2.5) defines the *rule of change of reference configuration* of the mechanical response. Notice that Eq. (2.5) permits us to define $W$ as a map on the space of 1-jets of (local) configurations which is independent on the chosen reference configuration. In fact, for each configuration $\phi$ we could define

$$W(j_{X,x}^1 \phi) = W(X, F),$$

where $F$ is the associated matrix to the 1-jet at $\phi_0(X)$ of $\phi \circ \phi_0^{-1}$.

It is remarkable that for any subbody $\mathcal{U}$ the mechanical response can be restricted to $\mathcal{U}$. So, the structure on the body induces a structure of elastic simple body over each subbody.

From now on we will refer to $\mathcal{B}$ simply as a *body*.

The locality of the mechanical response implies that we may talk about the material at each point of the body. In this sense, given two body points a natural question arises: when are they made of the same material? Let us assume that body $\mathcal{B}$ may be undergone to experimental observations under a fixed reference configuration $\phi_0$ and body points $X$ and $Y$ are made of the same material. Looking at them under the microscope we do not need to see exactly the same. It could happen that the arrangement surrounding one point is changed with respect to the other.

In our mathematical framework this means that the constitutive equation of one of them differs from the other only by an application of a linear transportation. These kinds of linear isomorphisms are called *material isomorphisms*.

**Definition 2.0.3.** Let $\mathcal{B}$ be a body. Two material particles $X, Y \in \mathcal{B}$ are said to be *materially isomorphic* if there exists a local diffeomorphism $\psi$ from an open neighborhood $\mathcal{U} \subseteq \mathcal{B}$ of $X$ to an open neighborhood $\mathcal{V} \subseteq \mathcal{B}$ of $Y$ such that $\psi(X) = Y$ and

$$W(X, F \cdot P) = W(Y, F), \qquad (2.6)$$

for all infinitesimal deformation $F$ where $P$ is given by the Jacobian matrix of $\phi_0 \circ \psi \circ \phi_0^{-1}$ at $\phi_0(X)$. The 1-jets of local diffeomorphisms satisfying Eq. (2.6) are called *material isomorphisms*. A material isomorphism from $X$ to itself is called a *material symmetry*. In cases where it causes no confusion we often refer to associated matrix $P$ as the material isomorphism (or symmetry).

Notice that, the identities at the vector spaces $T_X\mathcal{B}$ are obviously material isomorphisms. On the other hand, for any material isomorphism $P$ the inverse $P^{-1}$ is again a material isomorphism. In fact, by using Eq. (2.6)

$$W(X, F) = W(X, F \cdot P^{-1} \cdot P) = W(Y, F \cdot P^{-1}).$$

Finally, the composition preserves material isomorphisms. So, the relation of being "*materially isomorphic*" defines an equivalence relation (symmetric, reflexive and transitive) over the body manifold $\mathcal{B}$.

For any body point $X$ we denote by $G(X)$ the set of all material symmetries at $X$. Then, as a consequence we have that every $G(X)$ is a group. Therefore, it is trivial to prove that the material symmetry groups of materially isomorphic body points are conjugated, i.e., if $X$ and $Y$ are material isomorphic we have that

$$G(Y) = P \cdot G(X) \cdot P^{-1},$$

where $P$ is a material isomorphism from $X$ to $Y$.

**Proposition 2.0.4.** *Let $\mathcal{B}$ be a body. Two body points $X$ and $Y$ are materially isomorphic if and only if there exist two (local) configurations $\phi_1$ and $\phi_2$ such that*

$$W_1(X, F) = W_2(Y, F), \quad \forall F,$$

*where $W_i$ is the mechanical response associated to $\phi_i$ for $i = 1, 2$.*

**Proof.** Two body points $X$ and $Y$ are materially isomorphic if there exists a local diffeomorphism $\psi$ from $X$ to $Y$ such that

$$W(X, F \cdot P) = W(Y, F), \tag{2.7}$$

for all infinitesimal deformation $F$ where $P$ is given by the induced tangent map of $\phi_0 \circ \psi \circ \phi_0^{-1}$ at $\phi_0(X)$. Then, we define the local configuration $\phi_1 = \phi_0 \circ \psi$. Then, by using Eqs. (2.5) and (2.7), the local mechanical response $W_1$ induced by $\phi_1$ satisfies that

$$W_1(X, F) = W(X, F \cdot P) = W(Y, F). \qquad \square$$

This result gives us an intuition behind the notion of material isomorphism. In fact, two points will be made of the same material if the mechanical response is the same under the action of two (possibly different) reference configurations. Furthermore, as a corollary we have the following immediate result: *Condition of being materially isomorphic is configuration indifferent.*

**Definition 2.0.5.** A body $\mathcal{B}$ is said to be *uniform* if all of its body points are materially isomorphic.

Roughly speaking, a body is uniform if there are not two different materials inside the body. Notice that the definition of uniformity is a *pointwise* property. In fact, consider a uniform body $\mathcal{B}$ and a fixed body point $X_0$, for any other body point $Y$ we may choose a material isomorphism from $Y$ to $X_0$, say $P(Y) \in Gl(3, \mathbb{R})$. So, we can construct a map $P : \mathcal{B} \to Gl(3, \mathbb{R})$ consisting of material isomorphisms. Nevertheless, $P$ does not have to be differentiable. In other words, even when the body is uniform, the choice of the material isomorphisms is not, necessarily, smooth.

**Definition 2.0.6.** A body $\mathcal{B}$ is said to be *smoothly uniform* if for each point $X \in \mathcal{B}$ there is a neighborhood $\mathcal{U}$ around $X$ and a smooth map $P : \mathcal{U} \to Gl(3, \mathbb{R})$ such that for all $Y \in \mathcal{U}$ it satisfies that $P(Y)$ is a material isomorphism from $Y$ to $X$. The map $P$ is called a *left (local) smooth field of material isomorphisms*. A *right (local) smooth field of material isomorphisms* will be a smooth map $P : \mathcal{U} \to Gl(3, \mathbb{R})$ such that for all $Y \in \mathcal{U}$ it satisfies that $P(Y)$ is a material isomorphism from $X$ to $Y$.

Note that a given map $P : \mathcal{U} \to Gl(3, \mathbb{R})$ is a left smooth field of material isomorphisms if and only if the map $P^{-1} : \mathcal{U} \to Gl(3, \mathbb{R})$, such that $P^{-1}(Y)$ is the inverse of $P(Y)$, is a right smooth field of material isomorphisms. Hence, smooth uniformity may be equivalently characterized by using right smooth fields of material isomorphisms.

Assume that $P$ is a right (local) smooth field of material isomorphisms. Then, the mechanical response of the subbody $\mathcal{U}$ satisfies that

$$W(Y, F) = W(X, F \cdot P(Y)),$$

for all $Y \in \mathcal{U}$. Then, defining

$$\overline{W}(F) = W(X, F),$$

we have that

$$W(Y, F) = \overline{W}(F \cdot P(Y)). \tag{2.8}$$

The meaning of Eq. (2.8) is that the dependence of the mechanical response (near to a material particle) of the body coordinates is given by a multiplication of $F$ to the right by a right smooth field of material isomorphisms.

The following result shows that Eq. (2.8) defines a condition strictly weaker than the condition of being smoothly uniform.

**Proposition 2.0.7.** *Let $\mathcal{B}$ be a body. Then, suppose that the constitutive law $W$ of $\mathcal{B}$ respect to a reference configuration $\phi_0$ has associated a differentiable map $\overline{W} : Gl(3,\mathbb{R}) \to V$ satisfying Eq. (2.8) for a differentiable map $P : \mathcal{U} \to Gl(3,\mathbb{R})$. Then, $P$ is a smooth field of material isomorphisms if and only if*

$$\overline{W}(F) = \overline{W}(F \cdot P(X)). \qquad (2.9)$$

*where $X$ is a fixed point at the domain of $P$.*

**Proof.**    Assume that Eq. (2.9) is satisfied. Then, by Eq. (2.8)

$$\overline{W}(F \cdot P(X)) = W(X, F).$$

On the other hand, the same identity proves that for any $Y \in \mathcal{U}$

$$\overline{W}(F) = W(Y, F \cdot P(Y)^{-1}).$$

Hence, Eq. (2.9) implies that

$$W(X, F) = W(Y, F \cdot P(Y)^{-1}),$$

i.e., $P(Y)$ defines a material isomorphism from $X$ to $Y$. The converse is trivial.    □

Notice that condition Eq. (2.9) is not so strong. For instance, if the smooth fields $P$ go through the identity matrix, then Eq. (2.9) is immediately fulfilled.

Let $\mathcal{B}$ be a smoothly uniform body and $P : \mathcal{U} \to Gl(3,\mathbb{R})$ be a left (local) smooth field of material isomorphisms. Then, we have a tool to compare vectors at $\mathcal{U}$. In this sense, two tangent vectors $V_{X_1}$ and $V_{X_2}$ at two different material points $X_1$ and $X_2$ of $\mathcal{U}$ will be called *materially parallel with respect to $P$* if they have the same components by the action of $P$. In other words,

$$[P(X_2)^{-1} \circ P(X_1)](V_{X_1}) = V_{X_2}.$$

Here, for each $Y \in \mathcal{U}$ and $V_Y \in T_Y \mathcal{B}$, $P(Y)(V_Y)$ is given by the action of $P$ on the vector $V_Y$. In particular, considering $\psi^Y$ as a local diffeomorphism from $Y$ to $X$ such that $P(Y)$ is the associated matrix to $j^1_{Y,X}\psi^Y$ under the composition of the reference configuration, it satisfies that

$$P(Y)(V_Y) = T_Y\psi^Y(V_Y).$$

A vector field $\Theta \in \mathfrak{X}(\mathcal{U})$ is *materially constant with respect to* $P$ if for any two points $X_1$ and $X_2$ we have that $\Theta(X_1)$ and $\Theta(X_2)$ are materially parallel with respect to $P$. Equivalently, $\Theta$ is materially constant with respect to $P$ if and only if the vector

$$[P(Y)](\Theta(Y)),$$

does not depend on the choice of $Y \in \mathcal{U}$.

Let us express this condition locally. Let $(X^I)$ be the local coordinates generated by the reference configuration $\phi_0$. Then, for each $A$ we define the local vector field $P^A$ on $\mathcal{B}$ given by

$$P_A(Y) = P(Y)^{-1}\left(\frac{\partial}{\partial X^A_{|X}}\right).$$

Locally,

$$P_A = \left(P^{-1}\right)^B_A \frac{\partial}{\partial X^B}, \tag{2.10}$$

where,

$$\left(P^{-1}\right)^B_A(Y) = \frac{\partial(X^B \circ (\psi^Y)^{-1})}{\partial X^A_{|X}},$$

with $P(Y) = j^1_{Y,X}\psi^Y$. Then, $\Theta$ is materially constant with respect to $P$ if and only if $\Theta^B P^A_B$ is constant for all $A$, where $P^A_B$ is the inverse matrix to $\left(P^{-1}\right)^B_A$, i.e.,

$$P^B_A(Y) = \frac{\partial\left(X^B \circ \psi^Y\right)}{\partial X^A_{|Y}}.$$

Thus, $\Theta$ is materially constant if it satisfies that

$$P^A_B \frac{\partial \Theta^B}{\partial X^I} + \Theta^B \frac{\partial P^A_B}{\partial X^I} = 0, \quad \forall I$$

i.e.,

$$\frac{\partial \Theta^B}{\partial X^I} + \left(P^{-1}\right)^B_A \frac{\partial P^A_k}{\partial X^I}\Theta^k = 0, \quad \forall B, I \tag{2.11}$$

We could even find another characterization by using the so-called *material connections*. For a brief explanation on connections as covariant derivatives, see Appendix B. The material connection associated to $P$ is given by the unique covariant derivative $\nabla^P$ on $\mathcal{B}$ satisfying that

$$\nabla^P_{P_B} P_A = 0, \quad \forall A, B.$$

A straightforward calculation shows us that the *Christoffel symbols of* $\nabla^P$ are given by

$$\Gamma^K_{IJ} = \left(P^{-1}\right)^K_L \frac{\partial P^L_I}{\partial X^J}. \tag{2.12}$$

Therefore, $\Theta$ is materially constant with respect to $P$ if and only if $\nabla^P \Theta = 0$. It is important to note that the material connections are configuration-indifferent (see Jiménez *et al.*, 2019 or Chapter 5 for a proof).

Material connections have been masterfully treated by Wang (1967). In Wang and Truesdell (1973) various examples of material connections with non-vanishing curvature are studied. Other ways to construct material connections will be presented in Chapter 5 (Jiménez *et al.*, 2019).

We have defined the uniformity as the mathematical formalization of the following statement: *All the points are made of the same material.* Smooth uniformity consisted in a light imposition of smoothness on the body. A new more restrictive property is presented as the *local homogeneity of the body.*

In Noll's terminology, the notion of (local) homogeneity coincides with the absence of defects of the body. So, in a purely physical point of view, it is conceivable that the absence of defects may be fulfilled for non-uniform bodies. Homogeneity for non-uniform bodies has been properly defined in Epstein *et al.* (2019) and Jiménez *et al.* (2020) and will be dealt in this book in Chapter 6, Section 6.2.

The classical definition of (local) homogeneity means that the body may be depicted in such a configuration in which the translations of any point to any other are material isomorphisms. In other words, there exists a configuration which satisfies that all the points are indistinguishable from each other as far as the mechanical response concerned.

**Definition 2.0.8.** A body $\mathcal{B}$ is said to be *homogeneous* if it admits a global configuration $\phi$ which induces a left global smooth field of material isomorphisms $P$ such that $P(Y)$ is the associated matrix to the 1-jet

$$j^1_{Y,X}\left(\phi^{-1}\circ\tau_{\phi(X)-\phi(Y)}\circ\phi\right),\qquad(2.13)$$

via the reference configuration $\phi_0$, for all body point $Y\in\mathcal{B}$ and a fixed $X\in\mathcal{B}$ where $\tau_{\phi(X)-\phi(Y)}:\mathbb{R}^3\to\mathbb{R}^3$ denotes the translation on $\mathbb{R}^3$ by the vector $\phi(X)-\phi(Y)$. $\mathcal{B}$ is said to be *locally homogeneous* if there exists a covering of $\mathcal{B}$ by homogeneous open sets. $\mathcal{B}$ is said to be *(locally) inhomogeneous* if it is not (locally) homogeneous.

A left (local) smooth field of material isomorphisms $P$ is said to be *integrable* if it is induced by a (local) configuration $\phi$ via Eq. (2.13). These kinds of configuration are called *homogeneous configurations*. As a corollary of Proposition 2.0.7 we have the following result.

**Proposition 2.0.9.** *Let $\mathcal{B}$ be a body. Then, $\mathcal{B}$ is (locally) homogeneous if and only if there exist (local) reference configurations such that for the associated constitutive laws $W$ there are differentiable maps $\overline{W}:Gl(3,\mathbb{R})\to V$ satisfying that*

$$W(Y,F)=\overline{W}(F),$$

*for all body point $Y$ in the domain and $F\in Gl(3,\mathbb{R})$.*

Therefore, a material body is homogeneous if there exists a configuration such that the material response does not depend on the body points.

Let $\mathcal{B}$ be a (local) homogeneous body. Then, considering a homogeneous configuration as the reference configuration, the coordinates $\left(P^{-1}\right)^B_A$ of the (local) vector fields $P_A$ given in Eq. (2.10) satisfy

$$\left(P^{-1}\right)^B_A=\delta^B_A.$$

Then, the Christoffel symbols of the material connection $\nabla^P$ are zero respect to the homogeneous configuration. So, $P$ is integrable if there exists a local system of coordinates on $\mathcal{B}$ such that the partial derivatives are materially constant vector fields with respect to $P$.

Therefore, the material connection of $P$ provides a way to evaluate whether the smooth field of material isomorphisms $P$ is integrable: $P$ is integrable if and only if $\nabla^P$ is a flat and torsion-free covariant derivative (see Lemma B.0.13).

There is still another treatment of homogeneity by using the theory of $G$-structures (see Appendix C) in which the (local) homogeneity corresponds to the integrability of a particular $G$-structure. This approach can be found in Elżanowski *et al.* (1990) (see Elżanowski and Prishepionok, 1992 or Wang, 1967; see also Bloom, 1979; Maugin, 1993).

Let $\mathcal{B}$ be a smoothly uniform body. Fix $Z_0 \in \mathcal{B}$ and $\overline{Z}_0 = j^1_{0,Z_0}\phi \in F\mathcal{B}$ a frame at $Z_0$. Consider the following set:

$$\omega_{G_0}(\mathcal{B}) := \{j^1_{Z_0,Y}\psi \cdot \overline{Z}_0, \; : \; j^1_{Z_0,Y}\psi \text{ is a material isomorphism}\}.$$

Then, $\omega_{G_0}(\mathcal{B})$ is a $G_0$-structure on $\mathcal{B}$ (which contains $\overline{Z}_0$). In fact (see Proposition 4.0.47),

$$\omega_{G_0}(\mathcal{B}) = \Omega_{Z_0}(\mathcal{B}) \cdot \overline{Z}_0.$$

Notice that the structure group of $\omega_{G_0}(\mathcal{B})$ is given by

$$G_0 := \overline{Z}_0^{-1} \cdot G(Z_0) \cdot \overline{Z}_0.$$

A local section of $\omega_{G_0}(\mathcal{B})$ will be called *local uniform reference*. A global section of $\omega_{G_0}(\mathcal{B})$ will be called *global uniform reference*. We call *reference crystal* to any frame $\overline{Z}_0 \in F\mathcal{B}$ at $Z_0$.

**Remark 2.0.10.**

(1) If we change the point $Z_0$ to another body point $Z_1$ then we obtain an isomorphic $G_0$-structure. We only have to take a frame $\overline{Z}_1$ as the composition of $\overline{Z}_0$ with a material isomorphism $j^1_{Z_0,Z_1}\psi$.

(2) We have fixed a configuration $\phi_0$. Suppose that $\phi_1$ is another reference configuration such that the change of configuration is given by $\psi = \phi_1^{-1} \circ \phi_0$. Transporting the reference crystal $\overline{Z}_0$ via $F\psi$ we get another reference crystal such that the $G_0$-structures are isomorphic.

(3) Finally suppose that we have another crystal reference $\overline{Z}'_0$ at $Z_0$. Hence, the new $G'_0$-structure, $\omega_{G'_0}(\mathcal{B})$, is conjugate of $\omega_{G_0}(\mathcal{B})$, namely,

$$G'_0 = A \cdot G_0 \cdot A^{-1}, \quad \omega_{G'_0}(\mathcal{B}) = \omega_{G_0}(\mathcal{B}) \cdot A,$$

with $A = \overline{Z}'_0 \cdot \overline{Z}_0^{-1}$.

In this way, the definition of homogeneity in terms of $G$-structures is the following.

**Definition 2.0.11.** A body $\mathcal{B}$ is said to be *homogeneous* with respect to a given frame $\overline{Z}_0$ if it admits a global deformation $\kappa$ such that $\kappa^{-1}$ induces a uniform reference $P$, i.e., for each $X \in \mathcal{B}$

$$P(X) = j^1_{0,X}\left(\kappa^{-1} \circ \tau_{\kappa(X)}\right),$$

where $\tau_{\kappa(X)} : \mathbb{R}^3 \to \mathbb{R}^3$ denotes the translation on $\mathbb{R}^3$ by the vector $\kappa(X)$. $\mathcal{B}$ is said to be *locally homogeneous* if every $X \in \mathcal{B}$ has a neighborhood which is homogeneous.

This definition is equivalent to the first one. The proof is included in Chapter 5 in term of Lie groupoids (Jiménez *et al.*, 2019).

It is easy to prove the following result.

**Proposition 2.0.12.** *If $\mathcal{B}$ is homogeneous then $\omega_{G_0}(\mathcal{B})$ is integrable. Conversely, $\omega_{G_0}(\mathcal{B})$ is integrable implies that $\mathcal{B}$ is locally homogeneous.*

Notice that, using Remark 2.0.12, this result shows us that the homogeneity does not depend on the point, the reference configuration and the frame $\overline{Z}_0$.

Let us now present some examples of simple elastic bodies. More examples will be studied along the book.

**Solids.** Let $\mathcal{B}$ be a body with reference configuration $\phi_0$. A material particle $X$ is said to be *solid* if it satisfies that the group

$$G^X_0 = j^1_{X,0}\left(\tau_{-\phi_0(X)} \circ \phi_0\right) \cdot G(X) \cdot j^1_{0,X}\left(\phi_0^{-1} \circ \tau_{\phi_0(X)}\right),$$

called *material symmetry group of $X$ respect to $\phi_0$*, is a conjugated subgroup of a subgroup of the orthogonal group $\mathcal{O}$, where $j^1_{0,X}\left(\phi_0^{-1} \circ \tau_{\phi_0(X)}\right)$ is the linear frame on $\mathcal{B}$ induced by $j^1_{0,X}\left(\phi_0^{-1} \circ \tau_{\phi_0(X)}\right)$ (see Appendix C). $\mathcal{B}$ is said to be *solid* if all its body points are solid.

It is usual to assume that there exist (local) configurations $\phi_1$ such that the symmetry groups respect to $\phi_1$ are subgroups of $\mathcal{O}$ (*contorted aleotropy*).

Notice that, exactly, this is always true. These kinds of configurations are called *undistorted states*.

Thus, let $\mathcal{B}$ be a solid with a reference configuration $\phi_0$ which is an undistorted state. The pullback of the usual metric on $\mathbb{R}^3$ by $\phi_0$ results into a Riemannian metric $g_0$ on $\mathcal{B}$.

Let $j^1_{X,X}\psi$ be a material symmetry. Then, the associated matrix to

$$j^1_{0,0}\left(\tau_{-\phi_0(X)} \circ \phi_0 \circ \psi \circ \phi_0^{-1} \circ \tau_{\phi_0(X)}\right)$$

is an orthogonal matrix. In particular,

$$T_{\phi_0(X)}\left(\phi_0 \circ \psi \circ \phi_0^{-1}\right)(v) \cdot T_{\phi_0(X)}\left(\phi_0 \circ \psi \circ \phi_0^{-1}\right)(w) = v \cdot w,$$

for all $v, w \in \mathbb{R}^3 \cong T_{\phi_0(X)}\mathbb{R}^3$ where $\cdot$ is the scalar product in $\mathbb{R}^3$. Hence, by definition, for any two vector $V_X, W_X \in T_X\mathcal{B}$ we have that

$$\begin{aligned}
g_0(X)\,(&T_X\psi\,(V_X)\,, T_X\psi\,(W_X)) \\
&= T_X\,(\phi_0 \circ \psi)\,(V_X) \cdot T_X\,(\phi_0 \circ \psi)\,(W_X) \\
&= T_X\phi_0\,(V_X) \cdot T_X\phi_0\,(W_X) \\
&= g_0(X)(V_X, W_X).
\end{aligned}$$

In other words, the materials symmetries are isometries for the metric $g$.

Now, suppose that $\mathcal{B}$ *isotropic*, i.e., $G_0^X$ is a conjugated group of $\mathcal{O}$ for all body point $X$. In this case, the same argument proves that a local automorphism $\psi$ induces a material symmetry at a point $X$ if and only if $\psi$ induces an isometry at $X$ of $g_0$. So, there exist Riemannian metrics characterizing the material symmetries.

If the solid $\mathcal{B}$ is supposed to be smoothly uniform, we do not need to assume the existence of global (or local) undistorted states to construct these kinds of metrics.

Consider a left (local) smooth field of material isomorphisms $P$ around a material particle $X$ and a local configuration $\phi$ such that it is an undistorted state at $X$, i.e.,

$$G = j^1_{X,0}\left(\tau_{-\phi(X)} \circ \phi\right) \cdot G(X) \cdot j^1_{0,X}\left(\phi^{-1} \circ \tau_{\phi(X)}\right),$$

is a subgroup of $\mathcal{O}$. Notice that, exactly, the undistorted states always exist.

Then, we define a (local) Riemannian metric $g^P$ on $\mathcal{B}$ as follows:

$$g^P(Y)(V_Y, W_Y) = T_X\phi[P(Y)(V_Y)] \cdot T_X\phi[P(Y)(W_Y)]. \qquad (2.14)$$

We may assume (composing by the left with $P(X)^{-1}$) that $P(X)$ is the identity at $T_X\mathcal{B}$. Then, its satisfies that

$$g^P(Y)(V_Y, W_Y) = g^P(X)(P(Y)(V_Y), P(Y)(W_Y)). \qquad (2.15)$$

Otherwise speaking, the values of the metric $g^P$ is the combination of the values of the metric at a fixed point $X$ and the translations by $P$. Notice that, Eq. (2.15) shows us that the composition $T_X\phi \circ P$ is a *left (local) field of undistorted states*. In fact, for any material isomorphism $j^1_{Y,Z}\psi$ we have that the composition $P(Z) \circ T_Y\psi \circ P^{-1}(Y)$ defines an isometry for $g^P$. Here $P(Z)$ is being considered as the 1-jet $j^1_{Z,X}\psi^Z$ such that the associated matrix via the composition with the reference configuration is $P(Z)$.

So, the smooth uniformity permits us to extend differentiably any undistorted state at a fixed point to a field of undistorted states.

In this case, we have that the material isomorphisms are isometries of $g^P$.

Suppose that the Levy-Civita connection associated to $g^P$ is flat and torsion-free. Equivalently, there exists a system of coordinates $(y^i)$ on $\mathcal{B}$ such that $g_{ij} = g\left(\frac{\partial}{\partial y^i}, \frac{\partial}{\partial y^j}\right) = \delta^i_j$ (see, for instance, O'Neill, 1983). Then, by using Eq. (2.14) there exists a local chart $\varphi = (y^j)$ such that $T_X\phi \circ P(Y) \circ T_y\varphi^{-1}$ induces an orthogonal matrix for all $Y = \varphi^{-1}(y)$ in the domain. This fact implies ($P(X)$ is the identity) that

$$T_X\phi \circ T_x\varphi^{-1},$$

induces also an orthogonal matrix with $X = \varphi^{-1}(x)$. Therefore, $\varphi$ is indeed a local undistorted state. Thus, we have proved that the existence of local undistorted states is equivalent to that the Levi-Civita connection of $g^P$ is flat and torsion-free.

Assume now that the material body $\mathcal{B}$ is furthermore isotropic. Then, all the isometries are material isomorphisms and, hence, $\varphi$ induces a (local) smooth field of material isomorphisms via Eq. (2.13). Accordingly, $\mathcal{B}$ *is locally homogeneous if and only if the Levi-Civita connection of $g^P$ is flat and torsion-free.*

**Fluids.** Let $\mathcal{B}$ be a body with reference configuration $\phi_0$. A material particle $X$ is said to be *elastic fluid* if it satisfies that the material symmetry group of $X$ respect to $\phi_0$ is the unimodular group, i.e., the group of matrices with unit determinant. Notice that it does not make sense to take a conjugation of the unimodular group because any conjugation (by any non-singular matrix) of the unimodular group gives again the whole unimodular group. $\mathcal{B}$ is said to be *elastic fluid* if all its body points are elastic fluids.

Let $g_0$ be the Riemannian metric on $\mathcal{B}$ defined above and $V_0$ be its associated volume form. Then, it is easy to check that $j^1_{X,X}\psi$ is a material symmetry if and only if $\psi$ preserves $V_0$ at $X$, i.e.,

$$\psi^* V_0(X) = V_0(X).$$

Now, let us assume that $\mathcal{B}$ is smoothly uniform. Let $P$ be a left (local) smooth field of material isomorphisms $P$ around a material particle $X$. So, we may consider the metric $g^P$ defined above. Then, analogously to solids, we may prove that $j^1_{Y,Z}\psi$ is a material isomorphism if and only if $\psi$ preserves the volume form $V^P$ of $g^P$ at $Y$, i.e.,

$$\psi^* V^P(Y) = V^P(Z).$$

Therefore, immediately we have that any configuration induces a smooth field of material isomorphisms via Eq. (2.13). In other words, *any smoothly uniform elastic fluid is homogeneous.*

# Chapter 3

# Groupoids

In this chapter, we will study the notion of (*Lie*) *groupoid*. Groupoids are a natural generalization of groups and may be defined as particular kind of *categories*. While groupoids were presented in 1926 by Brandt (1927), categories were introduced later in 1945 by Eilenberg and MacLane (1945). In this sense, groupoids are defined as a "*small*" category such that every morphism is an isomorphism.

Adding differential structures we obtain the notion of *Lie groupoid* which was firstly introduced by Ehresmann in a collection of articles (Ehresmann, 1952, 1956, 1959, 1995) and redefined by Pradines (1966). Roughly speaking, a Lie groupoid is defined as a groupoid satisfying that the set of morphisms and the set of objects are differentiable manifolds and the *structure maps* are differentiable.

It is remarkable that, Continuum Mechanics is not the unique application of groupoids. In fact, (Lie) groupoids are useful tools in several mathematical areas, such as *Algebraic Topology*, *Differential Geometry*, *Galois Theory*, *Group Theory* or *Homotopy Theory* (see Brown, 1987; Ramsay and Renault, 2001). There are also several other research areas where groupoids are used such as *Geometric Mechanics* (Cortés *et al.*, 2006; de Diego and Sato Martín de Almagro, 2018; de León *et al.*, 2010; Ferraro *et al.*, 2017; Marrero *et al.*, 2006; Weinstein, 1996) and *Quantum Mechanics* (Ciaglia *et al.*, 2018). This book deals with another application of groupoids to *Theory of Elasticity*. It is, in fact, another contribution which extends the program proposed by Weinstein (1996) for *Continuum Mechanics*.

A good reference on groupoids is the famous book Mackenzie (2005). In Epstein (2010) and Weinstein (2001), we can find a more intuitive view of this topic. The book Valdés *et al.* (2006) (in Spanish) is also recommendable as a rigorous introduction to groupoids.

Let us start with some examples for a smooth presentation to the concept of groupoid.

**15-puzzle groupoid.** Sam Lloyd claimed in 1891 that he invented the *15-puzzle* (although some researches seem to prove that it is false). The popularity of this puzzle grew fastly (especially in Europe). The 15-puzzle consists of 15 little square blocks numbered from 1 to 15 next to an empty square enclosed in a $4 \times 4$ square box as it is shown in Fig. 3.1. The position of the squares exhibited in Fig. 3.1 is called *identity position*. Notice that the number of possible positions is exactly 16!

The permitted permutations of the puzzle are the sliding of the hole (one place at a time) in horizontal or vertical direction. Given any initial position, generally the goal of the game is to slide the squares around until you obtain a specified arrangement of the blocks (usually the identity position). In 1879 two American mathematicians W. W. Johnson and W. E. Story (Johnson and Story, 1879) achieved to prove that from any fixed initial position one cannot obtain any other random position. In fact, only half of all the possible positions can actually be obtained.

Observe that, mathematically speaking, the 15-puzzle is similar to *Rubik's Cube* because the goal is just to obtain a certain position by using only some kind of permutations. Nevertheless, as a difference with Rubik's cube, the permitted permutations depend on the

Fig. 3.1.   15-puzzle.

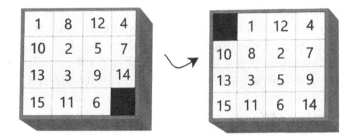

Fig. 3.2.   Transformation.

initial position. For instance, when the blank square is in a corner, it can only be moved towards two positions.

A *transformation* of the 15-puzzle is given by a sequence of permitted movements from a position to another. For example, in Fig. 3.2, we have a transformation resulting of moving the hole along the path $14 - 9 - 5 - 2 - 8 - 1$ from the initial position shown in this picture.

It is remarkable that any two transformations of the 15-puzzle cannot always be composed. In particular, the composition of two transformations can only be defined when the ending position of the first transformation is equal to the starting position of the second one. For this reason, unlike the set of transformations of the Rubik's cube, the set of transformations of the 15-puzzle does not have the structure of group. The structure of this set is the so-called *groupoid*.

Roughly speaking, the structure of groupoid is given by two sets,

- $\Gamma$ : *Set of transformations of the 15-puzzle*;
- $M$ : *Set of positions of the 15-puzzle*;

and a family of *structure maps* given by

- **Source and target maps**
  *Source and target maps* are given by two maps $\alpha, \beta : \Gamma \to M$ such that for any $g \in \Gamma$, $\alpha(g)$ (respectively $\beta(g)$) is the starting position (respectively the ending position) of the transformation $g$.
- **Section of identities**
  The *identity map* consists of a map $\epsilon : M \to \Gamma$ satisfying that for any position $x \in M$, $\epsilon(x)$ is the identity permutation of $x$, i.e., the blank square is not moved.

- **Inversion map**

  The *inversion map* is a map $i : \Gamma \to \Gamma$ where for each transformation $g \in \Gamma$, $i(g)$ is the opposite transformation. For example, for the transformation given in Fig. 3.2, the opposite transformation is simply 1–8–2–5–9–14 (exchanging the ending positions by the starting positions).

- **Composition law**

  By composing transformations we obtain a map $\cdot : \Gamma_{(2)} \to \Gamma$ where $\Gamma_{(2)}$ is just a subset of $\Gamma \times \Gamma$ given by the composable transformations.

These maps satisfy some properties such as the associativity of the composition which turn this structure into a groupoid called the 15-*puzzle groupoid*. The result of that for any two arbitrary positions generally there is not a transformation joining these two positions is translated in the language of groupoid as the 15-puzzle groupoid is not *transitive*.

**1-jets groupoid.** Fixing a manifold $M$ we consider the set, denoted by $\Pi^1(M, M)$, of all linear isomorphisms $L_{x,y} : T_x M \to T_y M$ for any $x, y \in M$. Any linear isomorphism $L_{x,y}$ has associated with it the points $x$ and $y$ of $M$. Denoting $x$ by $\alpha(L_{x,y})$ and $y$ by $\beta(L_{x,y})$ we can construct two maps $\alpha, \beta : \Pi^1(M, M) \to M$ which are the source and target maps respectively. Notice that the isomorphism $L_{x,y}$ can be composed with another element $G_{z,t}$ of $\Pi^1(M, M)$ if and only if

$$\alpha(G_{z,t}) = z = y = \beta(L_{x,y}).$$

So, as a difference with groups, the composition defines a partial multiplication on $\Pi^1(M, M)$. In fact, the domain of the multiplication is given by the set $\Pi^1(M, M)_{(2)}$ consisting of the elements $(G_{z,t}, L_{x,y}) \in \Pi^1(M, M) \times \Pi^1(M, M)$ such that $\alpha(G_{z,t}) = \beta(L_{x,y})$.

It is important to remark that the multiplication has similar properties as the multiplication of a group. Indeed, for each point $x \in M$ there exists the identity isomorphism $Id_{x,x} : T_x M \to T_x M$ which satisfy that

$$L_{x,y} Id_{x,x} = L_{x,y}, \quad Id_{x,x} G_{z,x} = G_{z,x},$$

for any two elements $L_{x,y}$ and $G_{z,x}$ of $\Pi^1(M, M)$ such that the above compositions are defined. Hence, the identities $Id_{x,x}$ generate the section of identities and act as unities for the partial multiplication in

$\Pi^1(M, M)$. Finally, any isomorphism $L_{x,y}$ has an inverse $L_{x,y}^{-1}$ satisfying that

$$L_{x,y}L_{x,y}^{-1} = Id_{y,y}, \quad L_{x,y}^{-1}L_{x,y} = Id_{x,x}.$$

This groupoid is called 1-*jets groupoid on M* and it will be properly studied in Examples 3.0.9 and 3.0.22. The 1-jets groupoid $\Pi^1(M, M)$ will have a great importance along the book.

The properties above presented can be written in a more abstract and rigorous way as follows:

**Definition 3.0.1.** Let $M$ be a set. A *groupoid* over $M$ is given by a set $\Gamma$ provided with the maps $\alpha, \beta : \Gamma \to M$ (*source map* and *target map* respectively), $\epsilon : M \to \Gamma$ (*section of identities*), $i : \Gamma \to \Gamma$ (*inversion map*) and $\cdot : \Gamma_{(2)} \to \Gamma$ (*composition law*) where for each $k \in \mathbb{N}$, $\Gamma_{(k)}$ is given by $k$ points $(g_1, \ldots, g_k) \in \Gamma \times \overset{k)}{\ldots} \times \Gamma$ such that $\alpha(g_i) = \beta(g_{i+1})$ for $i = 1, \ldots, k-1$. It satisfy the following properties:

(1) $\alpha$ and $\beta$ are surjective and for each $(g, h) \in \Gamma_{(2)}$,

$$\alpha(g \cdot h) = \alpha(h), \quad \beta(g \cdot h) = \beta(g).$$

(2) Associative law with the composition law, i.e.,

$$g \cdot (h \cdot k) = (g \cdot h) \cdot k, \quad \forall (g, h, k) \in \Gamma_{(3)}.$$

(3) For all $g \in \Gamma$,

$$g \cdot \epsilon(\alpha(g)) = g = \epsilon(\beta(g)) \cdot g.$$

In particular,

$$\alpha \circ \epsilon \circ \alpha = \alpha, \quad \beta \circ \epsilon \circ \beta = \beta.$$

Since $\alpha$ and $\beta$ are surjective we get

$$\alpha \circ \epsilon = Id_M, \quad \beta \circ \epsilon = Id_M,$$

where $Id_M$ is the identity at $M$.

(4) For each $g \in \Gamma$,

$$i(g) \cdot g = \epsilon(\alpha(g)), \quad g \cdot i(g) = \epsilon(\beta(g)).$$

Then,

$$\alpha \circ i = \beta, \quad \beta \circ i = \alpha.$$

These maps will be called *structure maps*. Furthermore, we will denote this groupoid by $\Gamma \rightrightarrows M$.

If $\Gamma$ is a groupoid over $M$, then $M$ is also denoted by $\Gamma_{(0)}$ and it is often identified with the set $\epsilon(M)$ of identity elements of $\Gamma$. $\Gamma$ is also denoted by $\Gamma_{(1)}$. Following the notation of the theory of categories (see Remark 3.0.3), the elements of $M$ are called *objects* and the elements of $\Gamma$ are called *morphisms*. The map $(\alpha, \beta) : \Gamma \to M \times M$ is called the *anchor map* and the space of sections of the anchor map is denoted by $\Gamma_{(\alpha,\beta)}(\Gamma)$. For any $g \in \Gamma$ the image $i(g)$ by the inversion map is denoted by $g^{-1}$.

Now, we define the morphisms of the category of groupoids.

**Definition 3.0.2.** If $\Gamma_1 \rightrightarrows M_1$ and $\Gamma_2 \rightrightarrows M_2$ are two groupoids then a *morphism of groupoids* from $\Gamma_1 \rightrightarrows M_1$ to $\Gamma_2 \rightrightarrows M_2$ consists of two maps $\Phi : \Gamma_1 \to \Gamma_2$ and $\phi : M_1 \to M_2$ such that for any $g_1 \in \Gamma_1$

$$\alpha_2(\Phi(g_1)) = \phi(\alpha_1(g_1)), \quad \beta_2(\Phi(g_1)) = \phi(\beta_1(g_1)), \qquad (3.1)$$

where $\alpha_i$ and $\beta_i$ are the source and the target map of $\Gamma_i \rightrightarrows M_i$ respectively, for $i = 1, 2$, and preserves the composition, i.e.,

$$\Phi(g_1 \cdot h_1) = \Phi(g_1) \cdot \Phi(h_1), \quad \forall (g_1, h_1) \in \Gamma_{(2)}.$$

We will denote this morphism as $\Phi$.

Observe that, as a consequence, $\Phi$ preserves the identities, i.e., denoting by $\epsilon_i$ the section of identities of $\Gamma_i \rightrightarrows M_i$ for $i = 1, 2$,

$$\Phi \circ \epsilon_1 = \epsilon_2 \circ \phi.$$

Then, using Eq. (3.1), $\phi$ is completely determined by $\Phi$. The category of groupoids will be denoted by $\mathcal{G}$.

Using this definition we define a *subgroupoid* of a groupoid $\Gamma \rightrightarrows M$ as a groupoid $\Gamma' \rightrightarrows M'$ such that $M' \subseteq M$, $\Gamma' \subseteq \Gamma$ and the inclusion map is a morphism of groupoids.

**Remark 3.0.3.** There is a more abstract way of defining a groupoid. We can say that a groupoid is a "small" category (the class of objects and the class of morphisms are sets) in which each morphism is invertible.

If $\Gamma \rightrightarrows M$ is the groupoid, then $M$ is the set of objects and $\Gamma$ is the set of morphisms.

A groupoid morphism is a functor between these categories which is a more natural definition.

We could even find another definition of groupoid given by Zakrzewski (1990a,b). Let $X, Y$ be two sets. A *relation* $r$ *from* $X$ *to* $Y$ is a triple $(Gr\,(r)\,, X, Y)$ with $Gr\,(r) \subseteq X \times Y$. A relation from $X$ to $Y$ will be denoted by $r : X \to Y$.

For a relation $r : X \to Y$ we will define its *transportation* $r^T : Y \to X$ by

$$(y, x) \in Gr\,(r^T) \leftrightarrow (x, y) \in Gr\,(r).$$

The *domain* of $r$ is the following set:

$$D\,(r) := \{x \in X : \exists y \in Y \ (x, y) \in Gr\,(r)\},$$

and the *image*,

$$Im\,(r) := \{y \in Y : \exists x \in X \ (x, y) \in Gr\,(r)\}.$$

A *composition* of relations $r : X \to Y$ and $s : Z \to X$ is the relation $rs : Z \to Y$ such that

$$Gr\,(rs) := \{(z, y) : \exists x \in X\,(z, x) \in Gr\,(s)\,, \ (x, y) \in Gr\,(r)\}.$$

Thus, given a family of sets $C$ we can form a category with these relations as morphisms and $C$ as the set of objects.

Cartesian product is defined in a natural way: Let $r_1 : X_1 \to Y_1$ and $r_2 : X_2 \to Y_2$ then $r_1 \times r_2 : X_1 \times X_2 \to Y_1 \times Y_2$ is given by the set $Gr\,(r_1 \times r_2)$ of elements $(x_1, x_2, y_1, y_2)$ such that $(x_1, y_1) \in Gr\,(r_1)$ and $(x_2, y_2) \in Gr\,(r_2)$.

Finally, a groupoid $\Gamma \rightrightarrows M$ is a quadruple $(\Gamma, m, i, \epsilon)$, where $\Gamma$ is a set, $m : \Gamma \times \Gamma \to \Gamma$, $\epsilon : \{1\} \to \Gamma$ (the symbol $\{1\}$ denotes a one point set) and $i : \Gamma \to \Gamma$ are relations such that

(i) $m\,(m \times Id) = m\,(Id \times m)$.
(ii) $m\,(\epsilon \times Id) = m\,(Id \times \epsilon) = Id$.
(iii) $i^2 = Id$.
(iv) Considering $\sigma : \Gamma \times \Gamma \to \Gamma \times \Gamma$ with $Gr\,(\sigma) := \{((x, y), (y, x)) : (x, y) \in \Gamma \times \Gamma\}$, it satisfies that

$$im = m\sigma\,(i \times i)\,.$$

(v) For all $\gamma \in \Gamma$,

$$\emptyset \neq Gr\,(m\,(i\,(\gamma)\,, \gamma)) \subseteq Im\,(\epsilon)\,.$$

Notice that $Id$ on a set $Z$ denotes the relation given by the diagonal $\Delta_Z \subset Z \times Z$.

Let us now present some examples of groupoids.

**Example 3.0.4.** A group is a groupoid over a point. In fact, let $G$ be a group and $e$ the identity element of $G$. Then, $G \rightrightarrows \{e\}$ is a groupoid, where the operation of the groupoid, $\cdot$, is the operation in $G$.

**Example 3.0.5.** Any set $X$ may be regarded as a groupoid on itself with $\alpha = \beta = \epsilon = i = Id_X$ and the operation on this groupoid is given by

$$x \cdot x = x, \quad \forall x \in X.$$

Note that, in this case, $X_{(2)} = \Delta_X$. We call this kind of groupoids as *base groupoids* and we will denote them as $\epsilon(X)$.

**Example 3.0.6.** For any set $A$ and any map $\pi : A \to M$, we can consider the pullback space $A \times_{\pi,\pi} A$ according to the diagram

i.e.,

$$A \times_{\pi,\pi} A := \{(a_x, b_x) \in A \times A / \pi(a_x) = x = \pi(b_x)\}.$$

Then, the maps,

$$\alpha(a_x, b_x) = a_x, \quad \beta(a_x, b_x) = b_x, \quad \forall (a_x, b_x) \in A \times_{\pi,\pi} A,$$
$$(c_x, b_x) \cdot (a_x, c_x) = (a_x, b_x), \quad \forall (c_x, b_x), (a_x, c_x) \in A \times_{\pi,\pi} A,$$
$$\epsilon(a_x) = (a_x, a_x), \quad \forall a_x \in A,$$
$$(a_x, b_x)^{-1} = (b_x, a_x), \quad \forall (a_x, b_x) \in A \times_{\pi,\pi} A$$

endow $A \times_{\pi,\pi} A$ with a structure of groupoid over $A$, called the *pair groupoid along* $\pi$. If $\pi = Id_A$ then this groupoid is called the *pair groupoid*.

Note that, if $\Gamma \rightrightarrows M$ is an arbitrary groupoid over $M$, then the anchor map $(\alpha, \beta) : \Gamma \to M \times M$ is a morphism from $\Gamma \rightrightarrows M$ to the pair groupoid of $M$.

The following example arises as a natural generalization of the previous one.

**Example 3.0.7.** Let $A, M$ be two sets, $A$ be a map $\pi : A \to M$ and $G$ be a group. Then we can construct a groupoid $A \times_{\pi,\pi} A \times G \rightrightarrows A$ where the source map is the second projection, the target map is the third projection and the composition law is given by the composition in $G$, i.e.,

$$(c_x, b_x, g) \cdot (a_x, c_x, h) = (a_x, b_x, g \cdot h),$$

for all $(c_x, b_x, g), (a_x, c_x, h) \in A \times_{\pi,\pi} A \times G$. This Lie groupoid is called *trivial groupoid along $\pi$ with group $G$*. When the map $\pi$ is the identity this groupoid is called *trivial groupoid on $A$ with group $G$*.

**Example 3.0.8.** Let $\pi : A \to M$ be a map and $\phi : G \times A \to A$ be a left action of a group $G$ on $A$ which preserves the fibres, i.e.,

$$\pi \circ \phi = \pi \circ pr_2,$$

where $pr_2 : G \times A \to A$ is the projection on the second coordinate.

We can construct the *transformation groupoid associated to $\phi$ along $\pi$* as follows:

- The set of morphisms is $G \times A$ and the set of objects is $A$.
- The source map and target map are given by

$$\alpha(g, a_x) = a_x, \quad \beta(g, a_x) = \phi(g, a_x),$$

for all $(g, a_x) \in G \times A$.
- The operation is

$$(g, \phi(h, a_x)) \cdot (h, a_x) = (gh, a_x),$$

for all $(h, a_x), (g, \phi(h, a_x)) \in G \times A$.
- The section of identities and inverse map are given by

$$\epsilon(a_x) = (e, a_x), \quad (g, a_x)^{-1} = \left(g^{-1}, \phi(g, a_x)\right),$$

for all $(g, a_x) \in G \times A$, where $e$ is the identity element in $G$.

It is easy to prove that $G \times A \rightrightarrows A$ is a groupoid which will be denoted by $G \ltimes_\pi A$. For a right action, we can define the transformation groupoid analogously and we will denote this groupoid by $G \rtimes_\pi A$. In the case in which $\pi$ is the identity map the groupoid is called the *transformation groupoid associated to* $\phi$ or simply the *transformation groupoid*.

Let us take a map $\pi : A \to M$ and a left action $\phi : G \times A \to A$ of a group $G$ on $A$ which preserves the fibres. Then, the map

$$\Phi : G \ltimes_\pi A \to G \times A \times_{\pi,\pi} A$$

$$(g, a_x) \mapsto (g, a_x, \phi(g, a_x))$$

is a morphism of groupoids. In fact, $\Phi$ is an isomorphism of Lie groupoids onto its image.

**Example 3.0.9.** Let $A$ be a vector bundle over a manifold $M$. Denote the set of all vector space isomorphisms $L_{x,y} : A_x \to A_y$ for $x, y \in M$, where for each $z \in M$ $A_z$ is the fibre of $A$ over $z$, by $\Phi(A)$. We can consider $\Phi(A)$ as a groupoid $\Phi(A) \rightrightarrows M$ such that, for all $x, y \in M$ and $L_{x,y} \in \Phi(A)$,

(i) $\alpha(L_{x,y}) = x$,
(ii) $\beta(L_{x,y}) = y$,
(iii) $L_{y,z} \cdot G_{x,y} = L_{y,z} \circ G_{x,y}$, $L_{y,z} : A_y \to A_z$, $G_{x,y} : A_x \to A_y$.

This groupoid is called the *frame groupoid on* $A$. As a particular case, when $A$ is the tangent bundle over $M$ we have the example $\Pi^1(M, M)$ introduced at the beginning of the chapter which is called *1-jets groupoid on* $M$. Notice that any isomorphism $L_{x,y} : T_x M \to T_y M$ can be written as a 1-jet $j^1_{x,y}\psi$ of a local diffeomorphism $\psi$ from $M$ to $M$ (to study the formalism of 1-jets see Appendix C).

Taking into account that any action can be seen as a particular groupoid (see Example 3.0.8), it makes sense to generalize the notions of orbit and isotropy group.

**Definition 3.0.10.** Let $\Gamma \rightrightarrows M$ be a groupoid with $\alpha$ and $\beta$ the source map and target map, respectively. For each $x \in M$, the set

$$\Gamma_x^x = \beta^{-1}(x) \cap \alpha^{-1}(x),$$

is called the *isotropy group of* $\Gamma$ at $x$. The set

$$\mathcal{O}(x) = \beta\left(\alpha^{-1}(x)\right) = \alpha(\beta^{-1}(x)),$$

is called the *orbit* of $x$, or *the orbit* of $\Gamma$ through $x$.

Notice that the orbit of a point $x$ consists of the points which are "*connected*" with $x$ by a morphism in the groupoid. It is also remarkable that inside the isotropy group the composition law is globally defined and, hence, it endows the isotropy groups with a *bona fide* group structure.

**Definition 3.0.11.** If $\mathcal{O}(x) = \{x\}$, or equivalently, $\beta^{-1}(x) = \alpha^{-1}(x) = \Gamma_x^x$ then $x$ is called a *fixed point*. *The orbit space of* $\Gamma$ is the space of orbits of $\Gamma$ on $M$, i.e., the quotient space of $M$ by the equivalence relation induced by $\Gamma$: two points of $M$ are equivalent if and only if they lie on the same orbit.

If $\mathcal{O}(x) = M$ for all $x \in M$, or equivalently $(\alpha, \beta) : \Gamma \to M \times M$ is a surjective map, the groupoid $\Gamma \rightrightarrows M$ is called *transitive*. If every $x \in M$ is fixed point, then the groupoid $\Gamma \rightrightarrows M$ is called *totally intransitive*. Furthermore, a subset $N$ of $M$ is called *invariant* if it is a union of some orbits.

Finally, the preimage of the source map $\alpha$ of a Lie groupoid at a point $x$ is called $\alpha$-*fibre at* $x$ and it is denoted by $\Gamma_x$. That of the target map $\beta$ is called $\beta$-*fibre at* $x$ and it is denoted by $\Gamma^x$.

**Definition 3.0.12.** Let $\Gamma \rightrightarrows M$ be a groupoid with $\alpha$ and $\beta$ the source and target map, respectively. We may define the left translation on $g \in \Gamma$ as the map $L_g : \Gamma^{\alpha(g)} \to \Gamma^{\beta(g)}$, given by

$$h \mapsto g \cdot h.$$

We may define the right translation on $g$, $R_g : \Gamma_{\beta(g)} \to \Gamma_{\alpha(g)}$ similarly.

Note that, the identity map on $\Gamma^x$ is given by

$$Id_{\Gamma^x} = L_{\epsilon(x)}. \tag{3.2}$$

So, for all $g \in \Gamma$, the left (respectively, right) translation on $g$, $L_g$ (respectively, $R_g$), is a bijective map with inverse $L_{g^{-1}}$ (respectively, $R_{g^{-1}}$).

Topological and differentiable structures could be imposed on a groupoid to get different kind of groupoids such as *topological groupoids* (see Valdés *et al.*, 2006). However, we will be mainly interested in *Lie groupoids*.

**Definition 3.0.13.** A *Lie groupoid* is a groupoid $\Gamma \rightrightarrows M$ such that $\Gamma$ is a smooth manifold, $M$ is a smooth manifold and all the structure maps are smooth. Furthermore, the source and the target map are submersions.

A *Lie groupoid morphism* is a groupoid morphism which is differentiable.

This definition permits us to construct the category of Lie groupoids, denoted by $\mathcal{LG}$, which is in fact a subcategory of the category $\mathcal{G}$ of groupoids.

**Definition 3.0.14.** Let $\Gamma \rightrightarrows M$ be a Lie groupoid. A *Lie subgroupoid* of $\Gamma \rightrightarrows M$ is a Lie groupoid $\Gamma' \rightrightarrows M'$ such that $\Gamma'$ and $M'$ are submanifolds of $\Gamma$ and $M$, respectively, and the inclusion maps $i_{\Gamma'} : \Gamma' \hookrightarrow \Gamma$ $i_{M'} : M' \hookrightarrow M$ become a morphism of Lie groupoids. $\Gamma' \rightrightarrows M'$ is said to be a *reduced Lie subgroupoid* if it is transitive and $M' = M$.

It is easy to check that if there exists a reduced Lie subgroupoid of a groupoid $\Gamma \rightrightarrows M$, then $\Gamma \rightrightarrows M$ is transitive.

Observe that, taking into account that $\alpha \circ \epsilon = Id_M = \beta \circ \epsilon$, then $\epsilon$ is an injective immersion.

On the other hand, in the case of a Lie groupoid, $L_g$ (respectively, $R_g$) is clearly a diffeomorphism for all $g \in \Gamma$.

Note also that, for each $k \in \mathbb{N}$, $\Gamma_{(k)}$ is a pullback space given by $\beta$ and the operation map on $\Gamma_{(k-1)}$. Thus, by induction, we may prove that $\Gamma_{(k)}$ is a smooth manifold for all $k \in \mathbb{N}$.

**Example 3.0.15.** A Lie group is a Lie groupoid over a point.

**Example 3.0.16.** Let $M$ be a smooth manifold, then the base groupoid $\epsilon(M)$ (see Example 3.0.5) is a Lie groupoid.

**Example 3.0.17.** Let $\pi : A \to M$ be a submersion. It is trivial to prove that the pair groupoid along $\pi$ is a Lie groupoid.

**Example 3.0.18.** Let $\pi : A \to M$ be a submersion and $G$ be a Lie group. Then, the trivial Lie groupoid along $\pi$ with group $G$, say $A \times_{\pi,\pi} A \times G \rightrightarrows A$, is obviously a Lie groupoid.

**Example 3.0.19.** Let $\pi : A \to M$ be a submersion and $\phi : G \times A \to A$ be a (left) action of a Lie group $G$ on $A$ which preserves the fibres. Then, the transformation groupoid $G \ltimes_\pi A$ associated to $\phi$ is a Lie groupoid.

**Example 3.0.20.** Let $\pi : P \to M$ be a principal bundle with structure group $G$. Denote by $\phi : G \times P \to P$ the action of $G$ on $P$.

Now, suppose that $\Gamma \rightrightarrows P$ is a Lie groupoid, with $\overline{\phi} : G \times \Gamma \to \Gamma$ a free and proper action of $G$ on $\Gamma$ such that, for each $g \in G$, the pair $(\overline{\phi}_g, \phi_g)$ is an isomorphism of Lie groupoids.

We can construct a Lie groupoid $\Gamma/G \rightrightarrows M$ such that the source map, $\overline{\alpha}$, and the target map, $\overline{\beta}$, are given by

$$\overline{\beta}([g]) = \pi(\beta(g)), \quad \overline{\alpha}([g]) = \pi(\alpha(g)),$$

for all $g \in \Gamma$, $\alpha$ and $\beta$ being the source and the target map on $\Gamma \rightrightarrows P$, respectively, and $[\cdot]$ denotes the equivalence class in the quotient space $\Gamma/G$. These kinds of Lie groupoids are called *quotient Lie groupoids by the action of a Lie group*.

There is an interesting particular case of the above example.

**Example 3.0.21.** Let $\pi : P \to M$ be a principal bundle with structure group $G$ and $P \times P \rightrightarrows P$ the pair groupoid. Take $\overline{\phi} : G \times (P \times P) \to P \times P$ the diagonal action of $\phi$, where $\phi : G \times P \to P$ is the action of $G$ on $P$.

Then it is easy to prove that $(\overline{\phi}_g, \phi_g)$ is an isomorphism of Lie groupoids and thus, we may construct the groupoid $(P \times P)/G \rightrightarrows M$. This groupoid is called *gauge groupoid* and is denoted by Gauge $(P)$.

**Example 3.0.22.** Let $A$ be a vector bundle over $M$ then the frame groupoid is a Lie groupoid (see Example 3.0.9). In fact, let $(x^i)$ and $(y^j)$ be local coordinate systems on open sets $U, V \subseteq M$ and $\{\alpha_p\}$ and $\{\beta_q\}$ be local basis of sections of $A_U$ and $A_V$, respectively. The corresponding local coordinates $(x^i \circ \pi, \alpha^p)$ and $(y^j \circ \pi, \beta^q)$ on $A_U$ and $A_V$ are as follows:

- For all $a \in A_U$,

$$a = \alpha^p(a)\alpha_p(x^i(\pi(a))).$$

- For all $a \in A_V$,

$$a = \beta^q(a)\beta_q(y^j(\pi(a))).$$

Then, we can consider a local coordinate system $\Phi(A)$

$$\Phi(A_{U,V}) : (x^i, y_i^j, y_i^j),$$

where, $A_{U,V} = \alpha^{-1}(U) \cap \beta^{-1}(V)$ and for each $L_{x,y} \in \alpha^{-1}(x) \cap \beta^{-1}(y) \subseteq \alpha^{-1}(U) \cap \beta^{-1}(V)$,

- $x^i(L_{x,y}) = x^i(x)$;
- $y^j(L_{x,y}) = y^j(y)$;
- $y_i^j(L_{x,y}) = A_{L_{x,y}}$, where $A_{L_{x,y}}$ is the associated matrix to the induced map of $L_{x,y}$ by the local coordinates $(x^i \circ \pi, \alpha^p)$ and $(y^j \circ \pi, \beta^q)$.

In particular, if $A = TM$, then the 1-jets groupoid on $M$, $\Pi^1(M, M)$, is a Lie groupoid and its local coordinates will be denoted as follows:

$$\Pi^1(U, V) : (x^i, y^j, y_i^j), \tag{3.3}$$

where, for each $j_{x,y}^1 \psi \in \Pi^1(U, V)$

- $x^i\left(j_{x,y}^1 \psi\right) = x^i(x)$.
- $y^j\left(j_{x,y}^1 \psi\right) = y^j(y)$.
- $y_i^j\left(j_{x,y}^1 \psi\right) = \dfrac{\partial\left(y^j \circ \psi\right)}{\partial x^i}\Big|_x$.

Let $\Gamma \rightrightarrows M$ be a Lie groupoid. Using that $\beta, \alpha$ are submersions, we have that the $\beta$-fibres and the $\alpha$-fibres are closed submanifolds of $\Gamma$. Moreover, the following lemma will be useful to prove some fundamental results on Lie groupoids.

**Lemma 3.0.23.** *Let $\phi : G \times M \to M$ be a free (left) action of a Lie group $G$ on a manifold $M$. The following conditions are equivalent.*

(i) *For any $x \in M$, there exists an embedded submanifold $N_x$ with $x \in N_x$ such that $G \times N_x \to M$ given by the restriction of the action of $G$ is an open embedding.*

(ii) *There is a smooth (not necessarily Hausdorff) structure on $M/G$ such that the quotient projection $M \to M/G$ is a principal $G$-bundle.*

(iii) *There exist a (perhaps non-Hausdorff) manifold $X$ and a smooth map $f : M \to X$ which is constant on the $G$-orbits and satisfies*

$$Ker\,(T_x f) = T_{(e,x)}\phi(\{0\} \times T_e G),$$

*for all $x \in M$.*

**Proof.** If (i) holds, then for each $x \in M$ the restriction of the quotient projection $N_x \to M/G$ is a topological open embedding (note that this map is trivially injective), and we may define a smooth structure on $M/G$ such that this map is a smooth open embedding. Therefore (i) implies (ii). Note that (iii) follows directly from (ii) ($X = M/G$ and $f$ is the quotient map). So we only need to prove that (iii) implies (i).

Take any $x \in M$, and choose a submersion $h : V \to \mathbb{R}^k$ defined on an open neighborhood $V$ of $f(x)$ in $X$ such that $Ker\,(T_{f(x)} h)$ is complementary to $T_x f\,(T_x M)$. Next, choose a small transversal section $N_x$ of the foliation of $M$ given by the connected components of the $G$-orbits, with $x \in N_x$ and $f(N_x) \subseteq V$. Now, by construction, $T_x\,(h \circ f_{|N_x})$ is an isomorphism, so we may shrink $N_x$ if necessary so that

$$h \circ f_{|N_x}$$

is an open embedding. In particular, $f$ is injective on $N_x$. Since $f$ is also constant along the $G$-orbits, it follows that each $G$-orbit intersects $N_x$ in at most one point. Since $N_x$ is transversal to the $G$-orbits, this proves $(i)$. $\square$

Thus, we may prove the following results.

**Lemma 3.0.24.** *If $\Gamma \rightrightarrows M$ is a Lie groupoid, then, for all $x, y \in M$, $\Gamma_x \cap \Gamma^y$ is a closed submanifold of $\Gamma$.*

**Proof.** First, we may construct the distribution $\mathcal{H}$ on $\Gamma$, given by

$$g \mapsto \mathcal{H}_g = Ker\,(T_g \alpha) \cap Ker\,(T_g \beta), \quad \forall g \in \Gamma.$$

Now, consider the left translation

$$L_g : \Gamma^{\alpha(g)} \to \Gamma^{\beta(g)},$$

which is a diffeomorphism between $\beta$-fibres. Observe that, for any $h \in \Gamma^{\alpha(g)}$, $\mathcal{H}_g$ is a subspace of $T_g \Gamma^{\alpha(g)} = Ker\,(T_g \beta)$. Using that

$\alpha \circ L_g = \alpha_{|\Gamma\alpha(g)}$, it follows that

$$T_{\epsilon(\alpha(g))} L_g \left( \mathcal{H}_{\epsilon(\alpha(g))} \right) = \mathcal{H}_g.$$

In addition, any basis $v_1, \ldots, v_k$ of $\mathcal{H}_{\epsilon(\alpha(g))}$ can be extended to a global frame $X_1, \ldots, X_k$ of $\mathcal{H}_{\Gamma\alpha(g)}$ by

$$X_i(g) = T_{\epsilon(\alpha(g))} L_g(v_i).$$

In this way, the restriction of $\mathcal{H}$ to the $\beta$-fibres is a locally finitely generated smooth distribution. It is involutive because it is exactly the kernel of the derivative of the map $\beta_{|\Gamma_{\alpha(g)}}$. Hence, using Hermann's Theorem A.0.22, it defines a foliation $\mathcal{F}_x$ of $\Gamma^x$ (which is parallelizable by the frame $X_1, \ldots, X_k$). The leaves of $\mathcal{F}_x$ are exactly the connected components of the $\alpha$-fibres of $\beta_{|\Gamma^x}$. So these fibres are closed manifolds.                                                                $\square$

Immediately we have the following corollary.

**Corollary 3.0.25.** *If $\Gamma \rightrightarrows M$ is a Lie groupoid, then for any $x \in M$, the isotropy group $\Gamma_x^x$ is a Lie group.*

Now, we can construct a left action of $\Gamma_x^x$ on $\beta^{-1}(x)$, $\phi : \Gamma_x^x \times \Gamma^x \to \Gamma^x$, given by

$$\phi(g, h) = L_g(h), \quad \forall (g, h) \in \Gamma_x^x \times \Gamma^x.$$

From this action, we can give structure of smooth manifold to the orbits as follows.

**Lemma 3.0.26.** *If $\Gamma \rightrightarrows M$ is a Lie groupoid, then for all $x \in M$ there is a natural structure of a smooth manifold on the orbit $\mathcal{O}(x)$ making $\alpha_{|\Gamma^x} : \Gamma^x \to \mathcal{O}(x)$ into a principal $\Gamma_x^x$-bundle*

**Proof.** As we have seen, the Lie group $\Gamma_x^x$ acts smoothly and freely on $\Gamma^x$ from the left, and it acts transitively along the manifolds $\Gamma_y \cap \Gamma^x$. Note that the condition (iii) of Lemma 3.0.23 is fulfilled by the map $\alpha_{|\Gamma^x}$, so the proposition implies that there is a natural structure of a smooth manifold on the orbit $\mathcal{O}(x)$ making $\alpha_{|\Gamma^x} : \Gamma^x \to \mathcal{O}(x)$ into a principal $\Gamma_x^x$-bundle. The fact that $M$ is Hausdorff implies that $\mathcal{O}(x)$ is also Hausdorff.                    $\square$

Observe that, taking into account that $\alpha_{|\Gamma^x} : \Gamma^x \to \mathcal{O}(x)$ is a principal $\Gamma^x_x$-bundle, we may consider the groupoid Gauge $(\Gamma^x)$ (see Example 3.0.21). So, as a corollary, we have the following result.

**Corollary 3.0.27.** *If* $\Gamma \rightrightarrows M$ *is transitive,* $\Gamma \cong$ Gauge $(\Gamma^x)$.

**Proof.** Consider the map

$$\Phi : \Gamma^x \times \Gamma^x/\Gamma^x_x \to \Gamma$$

$$[(g, h)] \mapsto g^{-1}h.$$

Suppose that $[(g, h)] = [(g', h')]$, then there exists $k \in \Gamma$ such that $g' = kg$ and $h' = kh$. Therefore,

$$\left(g'\right)^{-1} h' = (kg)^{-1}(kh) = g^{-1}h.$$

So, $\Phi$ is well defined. Furthermore, composing with the quotient projection map, we get that $\Phi$ is a smooth map.

Also, let $[(g, h)], [(g', h')] \in \Gamma^x \times \Gamma^x/\Gamma^x_x$ such that $\Phi([(g, h)]) = g^{-1}h = (g')^{-1} h' = \Phi([(g', h')])$. Then, taking $k = g'(g)^{-1}$, we have

$$kg = g', \quad kh = h',$$

i.e.,

$$[(g, h)] = [(g', h')].$$

On the other hand, let $k \in \Gamma$ with $\beta(k) = y$. Using that $\mathcal{O}(x) = M$, there exists $g \in \Gamma^x$ such that $\alpha(g) = y$. Hence, $(gk, g) \in \Gamma^x \times \Gamma^x$ and

$$\Phi([(g, gk)]) = k.$$

In this way, we have proved that $\Phi$ is a bijective map. So, it is clear that $\Phi$ is a Lie groupoid isomorphism over the identity. □

It is important to remark the importance of this result. In fact, we have proved that the only transitive Lie groupoids are the Gauge groupoids presented in Example 3.0.21.

# Chapter 4

# Algebroids

The notion of *Lie algebroid* was introduced by J. Pradines in 1966 (Pradines, 1966) as an infinitesimal version of Lie groupoid and for this reason the first name of this object was *infinitesimal groupoid*. To study this notion we also refer to Mackenzie (2005). Let us present a basic example of Lie algebroid to introduce the reader to the notion.

**Tangent bundle**

Let $M$ be a manifold. Then, the tangent bundle $TM$ of $M$ defines what is known as *Lie algebroid*. Consider the canonical projection $\pi_M : TM \to M$ of the tangent bundle of $M$. Then, the space of sections of $\pi_M$ is the module of vector fields $\mathfrak{X}(M)$ on $M$ and we have the following structure,

- **Anchor map:** The identity on $TM$ is a morphism of vector bundles from the domain of $\pi_M$, i.e. $TM$, to $TM$ which, in general, will be called the *anchor map*.
- **Lie bracket:** The Lie bracket $[\cdot, \cdot]$ of vector fields is a bracket on $\mathfrak{X}(M)$ such that $(\mathfrak{X}(M), [\cdot, \cdot])$ is a Lie algebra.
- **Leibniz rule:** It satisfies the following property,

$$[\Theta_1, f\Theta_2] = f[\Theta_1, \Theta_2] + \Theta_1(f)\Theta_2,$$

for all $\Theta_1, \Theta_2 \in \mathfrak{X}(M)$ and $f \in \mathcal{C}^\infty(M)$.

These properties turn the tangent bundle into a Lie algebroid. The definition of Lie algebroid will be properly introduced in Definition 4.0.1.

Consider now the pair Lie groupoid $M \times M \rightrightarrows M$ on $M$ (see Examples 3.0.6 and 3.0.17). A *left-invariant vector field* on the pair groupoid is simply a vector field $\Theta$ on $M \times M$ such that

$$\Theta\left(g \cdot h\right) = T_g L_h\left(\Theta\left(g\right)\right),$$

for all $g, h \in M \times M$ satisfying that $\alpha\left(g\right) = \beta\left(h\right)$. Notice that, by Definition 3.0.12, we have that $\Theta$ should be tangent to the $\beta$-fibres. Therefore, taking into account the left invariance, we have that the space of left-invariant vector fields $M \times M$ can be identified with the space of vector fields on $M$. This identification is, in fact, a Lie algebra morphism (the structure of Lie bracket is preserved).

Now, let $A\left(M \times M\right)$ be the vector bundle on $M$ such that the fibre $A\left(M \times M\right)_x$ at some $x \in M$ is given by the tangent space of the $\beta$-fibre at the identity morphism $\epsilon\left(x\right) = \left(x, x\right)$. This vector bundle will be what is called *the infinitesimal version of $M \times M$*.

Restricting the left-invariant vector fields on $M \times M$ to the identity morphism we obtain an isomorphism from the space of left-invariant vector fields on $M \times M$ to the space of sections $\Gamma\left(A\left(M \times M\right)\right)$ of $A\left(M \times M\right)$. This isomorphism endows the space $\Gamma\left(A\left(M \times M\right)\right)$ with a structure of Lie algebra which clearly satisfy the Leibniz rule. Hence, the infinitesimal version of the pair groupoid $M \times M \rightrightarrows M$ on $M$ will be (isomorphic to) the Lie algebroid structure of the tangent bundle on $M$.

Again, this construction will be explained with more rigorously along this chapter.

**Definition 4.0.1.** A *Lie algebroid over $M$* is a triple $(A \to M, \sharp, [\cdot, \cdot])$, where $\pi : A \to M$ is a vector bundle together with a vector bundle morphism $\sharp : A \to TM$, called the *anchor*, and a Lie bracket $[\cdot, \cdot]$ on the space of sections, such that the Leibniz rule holds

$$[\Lambda_1, f\Lambda_2] = f\left[\Lambda_1, \Lambda_2\right] + \sharp\left(\Lambda_1\right)\left(f\right)\Lambda_2, \tag{4.1}$$

for all $\Lambda_1, \Lambda_2 \in \Gamma\left(A\right)$ and $f \in \mathcal{C}^\infty\left(M\right)$.

$A$ is *transitive* if $\sharp$ is surjective and *totally intransitive* if $\sharp \equiv 0$. Also, $A$ is said to be *regular* if $\sharp$ has constant rank.

Looking at $\sharp$ as a $\mathcal{C}^\infty\,(M)$-module morphism from $\Gamma\,(A)$ to $\mathfrak{X}\,(M)$, for each section $\Lambda_1 \in \Gamma\,(A)$ we are going to denote $\sharp\,(\Lambda_1)$ by $\Lambda_1^\sharp$. Next, let us show the following fundamental property:

**Lemma 4.0.2.** *If $(A \to M, \sharp, [\cdot, \cdot])$ is a Lie algebroid, then the anchor map is a morphism of Lie algebras, i.e.*

$$[\Lambda_1, \Lambda_2]^\sharp = \left[\Lambda_1^\sharp, \Lambda_2^\sharp\right], \quad \forall \Lambda_1, \Lambda_2 \in \Gamma(A). \tag{4.2}$$

**Proof.** Let $\Lambda_1, \Lambda_2 \in \Gamma\,(A)$. By the Jacobi identity, for any section $\gamma \in \Gamma\,(A)$ and any function $f \in \mathcal{C}^\infty\,(M)$, we have

$$0 = [[\Lambda_1, \Lambda_2], f\gamma] + [[f\gamma, \Lambda_1], \Lambda_2] + [[\Lambda_2, f\gamma], \Lambda_1]. \tag{4.3}$$

Now, using the Leibniz rule,

- $[[\Lambda_1, \Lambda_2], f\gamma] = f\,[[\Lambda_1, \Lambda_2], \gamma] + [\Lambda_1, \Lambda_2]^\sharp\,(f)\,\gamma.$
- $[[f\gamma, \Lambda_1], \Lambda_2] = f\,[\Lambda_2, [\Lambda_1, \gamma]] + \Lambda_2^\sharp\,(f)\,[\Lambda_1, \gamma] + \Lambda_1^\sharp\,(f)\,[\Lambda_2, \gamma] + \Lambda_2^\sharp\left(\Lambda_1^\sharp\,(f)\right)\gamma.$
- $[[\Lambda_2, f\gamma], \Lambda_1] = -f\,[\Lambda_1, [\Lambda_2, \gamma]] - \Lambda_1^\sharp\,(f)\,[\Lambda_2, \gamma] - \Lambda_2^\sharp\,(f)\,[\Lambda_1, \gamma] - \Lambda_1^\sharp\left(\Lambda_2^\sharp\,(f)\right)\gamma.$

If we replace these equalities in Eq. (4.3), we have

$$0 = [\Lambda_1, \Lambda_2]^\sharp\,(f)\,\gamma + \Lambda_2^\sharp(\Lambda_1^\sharp\,(f))\gamma - \Lambda_1^\sharp(\Lambda_2^\sharp\,(f))\gamma$$

$$= [\Lambda_1, \Lambda_2]^\sharp\,(f)\,\gamma - \left[\Lambda_2^\sharp, \Lambda_1^\sharp\right](f)\,\gamma$$

for any $\gamma \in \Gamma\,(A)$ and for any $f \in \mathcal{C}^\infty\,(M)$. Thus, we conclude that

$$[\Lambda_1, \Lambda_2]^\sharp = [\Lambda_1^\sharp, \Lambda_2^\sharp], \quad \forall \Lambda_1, \Lambda_2 \in \Gamma(A). \qquad \square$$

**Remark 4.0.3.** Equation (4.2) is often considered as a part of the definition of a Lie algebroid though, as we have seen, it is a consequence of the other conditions.

Let $x$ be a point at $M$ and $A_x$ be the fibre of the Lie algebroid $A$ at $x$. Then, we may define a linear map $\sharp_x : A_x \to T_x M$ as the restriction of the anchor $\sharp$ to the fibres $A_x$ and $T_x M$.

**Definition 4.0.4.** The *isotropy algebra* of the Lie algebroid $A$ at the point $x \in M$ is the Lie algebra $(\mathrm{Ker}\,(\sharp_x), [\cdot, \cdot]_x)$, the Lie bracket is given by

$$[\Lambda_{1x}, \Lambda_{2x}]_x = [\Lambda_1, \Lambda_2](x),$$

for any two sections $\Lambda_1, \Lambda_2 \in \Gamma(A)$ such that

$$\Lambda_i(x) = \Lambda_{ix}, \quad i = 1, 2.$$

It is not hard to prove (see Example 4.0.6) that the bracket $[\cdot, \cdot]_x$ is well defined and, hence, induces a Lie algebra structure on the vector space $\mathrm{Ker}\,(\sharp_x)$.

An important remark is that the Lie algebra structure on sections is of local type, i.e. $[\Lambda_1, \Lambda_2](x)$ will depend on $\Lambda_2$ (therefore, on $\Lambda_1$ too) around $x$ only, $\forall x \in M$. Indeed, if $\Lambda_2, \widehat{\Lambda_2} \in \Gamma(A)$ with

$$\Lambda_{2|U} = \widehat{\Lambda_{2}}_{|U},$$

for an open neighborhood $U$ of $x$, taking $f \in \mathcal{C}^\infty(M)$ such that $\mathrm{supp}\,(f) \subseteq U$, $f \equiv 1$ on a compact neighborhood $V_x \subset U$ of $x$, then $f\Lambda_2 = f\widehat{\Lambda_2}$ on $M$. Using the Leibniz rule

$$[\Lambda_1, f\Lambda_2](x) = [\Lambda_1, \Lambda_2](x) + \Lambda_1^\sharp(x)(f)\Lambda_2(x)$$
$$= [\Lambda_1, \Lambda_2](x),$$

since $\Lambda_1^\sharp(x)(f) = 0$ ($f$ is constant on a neighborhood of $x$). Thus,

$$\left[\Lambda_1, \widehat{\Lambda_2}\right](x) = \left[\Lambda_1, f\widehat{\Lambda_2}\right](x)$$
$$= [\Lambda_1, f\Lambda_2](x)$$
$$= [\Lambda_1, \Lambda_2](x).$$

Finally, from skew-symmetry the result is proved.

As a consequence, the restriction of a Lie algebroid over $M$ to an open subset of $M$ is again a Lie algebroid. Taking local coordinates $(x^i)$ on $M$ and a local basis of sections of $A$, $\{\Lambda_p\}$, the corresponding local coordinates $(x^i \circ \pi, \Lambda^p)$ on $A$, satisfy

$$a = \Lambda^p(a)\Lambda_p\left(x^i(\pi(a))\right), \quad \forall a \in \pi^{-1}(U).$$

Such coordinates determine local functions $\sharp_p^i$, $C_{pq}^r$ on $M$ which contain the local information of the Lie algebroid structure, and accordingly they are called *the structure functions of the Lie algebroid*. They are given by

$$\Lambda_p^\sharp = \sharp_p^i \frac{\partial}{\partial x^i},$$

$$[\Lambda_p, \Lambda_q] = C_{pq}^r \Lambda_r.$$

Imposing Eq. (4.2) and the Jacobi identity over the local basis $\{\Lambda_p\}$, we get the following equations

$$C_{pq}^r \sharp_r^i = \left( \sharp_p^r \frac{\partial \sharp_q^i}{\partial x^r} - \sharp_q^r \frac{\partial \sharp_p^i}{\partial x^r} \right),$$

$$\sum_{\text{cyc}} \sharp_p^t \frac{\partial C_{qr}^s}{\partial x^t} + C_{pt}^s C_{qr}^t = 0,$$

for all $i, p, q$, where $\sum_{\text{cyc}} a_{ijk}$ means the cyclic sum $a_{ijk} + a_{kij} + a_{jki}$. These equations are usually called *structure equations*.

Now, we will give some examples of Lie algebroids.

**Example 4.0.5.** Any Lie algebra is a Lie algebroid over a single point. Indeed, identifying $\Gamma(\mathfrak{g})$ with $\mathfrak{g}$, the Lie bracket on sections is simply the Lie algebra bracket and the anchor map is the trivial one.

This kind of Lie algebroid is a particular case of the following example.

**Example 4.0.6.** Let $(A \to M, \sharp, [\cdot, \cdot])$ be a Lie algebroid where $\sharp \equiv 0$. Then, the Lie bracket on $\Gamma(A)$ is a point-wise Lie bracket, that is, the restriction of $[\cdot, \cdot]$ to the fibres induces a Lie algebra structure on each of them. More precisely, using that $\sharp \equiv 0$, the Leibniz rule is just the $\mathcal{C}^\infty(M)$-linearity, i.e.,

$$[\Lambda_1, f\Lambda_2] = f[\Lambda_1, \Lambda_2], \quad \forall f \in \mathcal{C}^\infty(M), \quad \forall \Lambda_1, \Lambda_2 \in \Gamma(A). \quad (4.4)$$

Consider $x \in M$ and $\Lambda_2, \widehat{\Lambda_2} \in \Gamma(A)$ such that

$$\Lambda_2(x) = \widehat{\Lambda_2}(x).$$

Let $\{\gamma_1, \ldots, \gamma_k\}$ be a basis of local sections. Then, around $x$, we have

$$\Lambda_2 - \widehat{\Lambda_2} = f_i \gamma_i,$$

with $f_i(x) = 0$, for all $i \in \{1, \dots, k\}$. From Eq. (4.4) for each $\Lambda_1 \in \Gamma(A)$ we have

$$[\Lambda_1, \Lambda_2 - \widehat{\Lambda_2}](x) = f_i(x)[\Lambda_1, \gamma_i](x) = 0,$$

Therefore,

$$[\Lambda_1, \Lambda_2](x) = [\Lambda_1, \widehat{\Lambda_2}](x).$$

Finally, skew-symmetry allows us to prove that the value of $[\Lambda_1, \Lambda_2]$ in a point $x \in M$ depends only on $\Lambda_1(x)$ and $\Lambda_2(x)$. These kinds of Lie algebroids (with $\sharp \equiv 0$) are called *Lie algebra bundles*. Note that the Lie algebra structures on the fibres are not necessary isomorphic to each other.

**Example 4.0.7.** Following the initial example, for any smooth manifold $M$, the tangent bundle of $M$, $TM$, is a Lie algebroid: the anchor map is the identity map and the Lie bracket is the usual Lie bracket of vector fields. This is called the *tangent algebroid* of $M$.

**Example 4.0.8.** Let $M$ be a manifold and $\mathfrak{g}$ be a Lie algebra. We can construct a Lie algebroid structure over the vector bundle $A = TM \oplus (M \times \mathfrak{g}) \to M$ such that

(i)  the anchor $\sharp : TM \oplus (M \times \mathfrak{g}) \to TM$ is the projection;
(ii) Lie algebra structure over the space of sections is given by

$$[X \oplus f, Y \oplus g] = [X, Y] \oplus \{X(g) - Y(f) + [f, g]\},$$

for all $X \oplus f, Y \oplus g \in \Gamma(A)$.

This Lie algebroid is called the *trivial Lie algebroid on M with structure algebra* $\mathfrak{g}$.

**Example 4.0.9.** If $M$ is a manifold and $D$ is an involutive subbundle of $TM$, then $D$ is a Lie algebroid over $M$, where the anchor is the inclusion $i : D \to TM$ and the bracket is the restriction of the Lie bracket of vector fields. Thus, let $\mathcal{F}$ be a regular foliation of $M$. Then the *tangent algebroid* of $\mathcal{F}$ is the subbundle of $TM$, $T\mathcal{F}$, consisting of tangent spaces to $\mathcal{F}$ with the usual Lie bracket, and the inclusion map as the anchor.

Note that, since $\mathcal{F}$ is regular, $T\mathcal{F}$ is a subbundle of $TM$, and its sections are vector fields tangent to $\mathcal{F}$. Moreover, $T\mathcal{F}$ being regular and integrable, implies that it is involutive and, as a consequence, the Lie bracket of two vector fields tangent to $\mathcal{F}$ is again a vector field tangent to $\mathcal{F}$.

**Example 4.0.10.** Let $M$ be a smooth manifold, $\mathfrak{g}$ be a Lie algebra and $\xi : \mathfrak{g} \to \mathfrak{X}(M)$ be a Lie algebra morphism (i.e. $\xi : \mathfrak{g} \to \mathfrak{X}(M)$ is an infinitesimal action of a Lie algebra $\mathfrak{g}$ on a manifold $M$). It is possible to associate to it the following *transformation algebroid*:

- **Vector bundle:** The vector bundle is the trivial bundle $\mathfrak{g} \times M \to M$.
- **Anchor map:** The anchor map is $\sharp : \mathfrak{g} \times M \to TM$ such that

$$\sharp(u, x) = \xi(u)(x).$$

So, the anchor map is the fixed Lie algebra morphism $\xi$.
- **Lie bracket:** The Lie bracket is given by

$$[\Lambda_1, \Lambda_2](z) = [\Lambda_1(z), \Lambda_2(z)]_{\mathfrak{g}} + (\xi(\Lambda_1(z)))_z(\Lambda_2)$$
$$- (\xi(\Lambda_2(z)))_z(\Lambda_1), \tag{4.5}$$

where we are identifying $\Lambda_1 \in \Gamma(\mathfrak{g} \times M)$ with a smooth map $\Lambda_1 : M \to \mathfrak{g}$.

In particular, if $\Lambda_1$ and $\Lambda_2$ are two constant sections then their bracket is a constant section given by the Lie bracket on $\mathfrak{g}$. Note that the last two terms in Eq. (4.5) are due to the Leibniz rule. We will denote the transformation algebroid of an action of $\mathfrak{g}$ on $M$ by $\mathfrak{g} \ltimes M$.

**Example 4.0.11.** Let $(M, \omega)$ be a pair where $M$ is a smooth manifold and $\omega \in \Omega^2(M)$ is a closed 2-form on $M$. Consider the vector bundle $A = TM \oplus (M \times \mathbb{R}) \to M$. Then, we may define the map

$$\sharp : A \to TM$$

$$u_x \oplus (x, t) \mapsto u_x.$$

In addition, note that the space $\Gamma(A)$ can be identified with the space

$$\overline{\Gamma(A)} := \{X \oplus f : X \in \mathfrak{X}(M), \ f \in \mathcal{C}^{\infty}(M)\}.$$

So, we construct a bracket on $\Gamma(A)$ characterized by

$$[X \oplus f, Y \oplus g] = [X, Y] \oplus (X(g) - Y(f) + \omega(X, Y)),$$

for all $X \oplus f, Y \oplus g \in \overline{\Gamma(A)}$. These maps define a Lie algebroid structure on $A$ which is transitive. In fact, the Jacobi identity is equivalent to the fact that $\omega$ is closed. Notice the similarity of this example with Example 4.0.8. In fact, roughly speaking, this example may be considered as in the trivial Lie algebroid on $M$ with structure algebra $\mathbb{R}$ (where the algebra structure on $\mathbb{R}$ is $[\cdot, \cdot] \equiv 0$) perturbed with a closed 2-form $\omega$.

**Example 4.0.12.** Let $\tau : P \to M$ be a principal bundle with structure group $G$. Denote by $\phi : G \times P \to P$ the action of $G$ on $P$. Now, suppose that $(A \to P, \sharp, [\cdot, \cdot])$ is a Lie algebroid, with vector bundle projection $\pi : A \to P$ and that $\overline{\phi} : G \times A \to A$ is an action of $G$ on $A$ such that $\pi$ is a vector bundle action under the action $\overline{\phi}$ where for each $g \in G$, the pair $(\overline{\phi}_g, \phi_g)$ satisfies that

(1) $\sharp \circ \overline{\phi}_g = T\phi_g \circ \sharp$;
(2) $\left[\overline{\phi}_g \circ \Lambda_1 \circ \phi_g^{-1}, \overline{\phi}_g \circ \Lambda_2 \circ \phi_g^{-1}\right] = \overline{\phi}_g \circ [\Lambda_1, \Lambda_2] \circ \phi_g^{-1}, \; \forall \Lambda_1, \Lambda_2 \in \Gamma(A)$.

This fact will be equivalent to the fact that $(\overline{\phi}_g, \phi_g)$ is a Lie algebroid isomorphism. Let $\overline{\pi} : A/G \to M$ be the quotient vector bundle of $\pi$ by the action of $G$. Then, we are going to construct a Lie algebroid structure on $\overline{\pi}$.

Denote by $\overline{\tau} : A \to A/G$ the quotient projection. Then, we may define the anchor map $\overline{\sharp} : A/G \to TM$ by

$$\overline{\sharp}(u) = T_{\pi(a)}\tau(\sharp(a)),$$

for all $u \in A/G$ and $a \in A$, where $\overline{\tau}(a) = u$.

Let $a, b \in A$ such that $\overline{\tau}(a) = \overline{\tau}(b) = u$. Then, there exists $g \in G$ such that

$$\overline{\phi}_g(b) = a.$$

Thus, since $\sharp \circ \overline{\phi}_g = T\phi_g \circ \sharp$, we have

$$\sharp(a) = \sharp\left(\overline{\phi}_g(b)\right) = T_{\tau(b)}\phi_g(\sharp(b)),$$

and therefore

$$T_{\pi(a)}\tau\left(\sharp\left(a\right)\right) = \left\{T_{\pi(b)}\left(\tau\circ\phi_g\right)\right\}\left(\sharp\left(b\right)\right) = T_{\pi(b)}\tau\left(\sharp\left(b\right)\right),$$

i.e., $\bar{\sharp}$ is well defined.

Furthermore, by construction

$$\bar{\sharp}\circ\bar{\tau} = T\tau\circ\sharp.$$

So, using that $\bar{\tau}$ is a submersion, the anchor is a smooth map. Finally, it is trivial to prove that $\bar{\sharp}$ is a vector bundle morphism.

On the other hand, for each $\Lambda_1, \Lambda_2 \in \Gamma\left(A\right)^G$ and for each $g \in G$

$$\overline{\phi}_g\circ\Lambda_1\circ\phi_{g^{-1}} = \Lambda_1, \quad \overline{\phi}_g\circ\Lambda_2\circ\phi_{g^{-1}} = \Lambda_2.$$

Then,

$$[\Lambda_1, \Lambda_2] = \left[\overline{\phi}_g\circ\Lambda_1\circ\phi_{g^{-1}}, \overline{\phi}_g\circ\Lambda_2\circ\phi_{g^{-1}}\right].$$

Using (2), we have

$$[\Lambda_1, \Lambda_2] = \overline{\phi}_g\circ[\Lambda_1, \Lambda_2]\circ\phi_{g^{-1}},$$

i.e., $[\Lambda_1, \Lambda_2] \in \Gamma\left(A\right)^G$. As a consequence, the Lie bracket on $\Gamma\left(A\right)$ restricts to $\Gamma\left(A\right)^G \cong \Gamma\left(A/G\right)$ and then, this structure induces a Lie algebra structure on $\Gamma\left(A/G\right)$. Finally, it is easy to prove that the Leibniz identity is satisfied. This kind of Lie algebroids is called *quotient Lie algebroid by the action of a Lie group*.

A particular but interesting example of this construction is obtained when we consider the tangent lift of a free and proper action of a Lie group on a manifold.

**Example 4.0.13.** Let $\pi : P \to M$ be a principal bundle with structure group $G$. Denote by $\phi$ the (left) action of $G$ on $P$. Let $(TP \to P, Id_{TP}, [\cdot, \cdot])$ be the tangent algebroid and $\phi^T : G \times TP \to TP$ be the tangent lift of $\phi$.

Then, $\phi^T$ satisfies the conditions of Example 4.0.12. Thus, one may consider the quotient Lie algebroid $\left(TP/G \to M, \bar{\sharp}, \overline{[\cdot, \cdot]}\right)$ by the action of $G$. This algebroid is called the *Atiyah algebroid associated with the principal bundle* $\pi : P \to M$.

Note that, as we have seen, the space of sections can be considered as the space of invariant vector field by the action $\phi$ over $M$.

Next, we introduce the definition of a morphism in the category of Lie algebroids. However, the case of Lie algebroids is not as easy as in the case of groupoids. The difficulty lies on the fact that, in general, a morphism between vector bundles does not induce a map between the modules of sections. This implies that a relation between the brackets of the space of sections from a morphism of vector bundles is not immediately clear.

The definition of morphism of Lie algebroids was introduced by Pradines (1967). Nevertheless, that definition was not simple enough to be used. In the article Almeida and Kumpera (1981) the authors give another, more conceptual, definition. Even in that case, the difficulties do not disappear at all and it is still difficult to work with it. It was necessary another definition (obtained from an observation made by Weinstein to Mackenzie about the Lie algebroid of an action groupoid) to solve this.

We will show a direct definition in terms of $(\Phi, \phi)$-decompositions of sections which is easy to understand.

**Definition 4.0.14.** Let $\pi : A \to M$ and $\pi' : A' \to M'$ be vector bundles and $(\Phi, \phi)$, with $\Phi : A' \to A$ and $\phi : M' \to M$ a vector bundle morphism. If $\Lambda \in \Gamma(A)$ and $\sigma \in \Gamma(A')$ satisfy

$$\Phi \circ \sigma = \Lambda \circ \phi,$$

then we say that $\sigma$ and $\Lambda$ are $(\Phi, \phi)$-related and we write $\sigma \sim_{(\Phi,\phi)} \Lambda$.

We also say that $\sigma \in \Gamma(A')$ is $(\Phi, \phi)$-projectable if it is $(\Phi, \phi)$-related to some $\Lambda \in \Gamma(A)$.

It is easy to prove that this relation is $\mathcal{C}^\infty(M)$-linear in the sense that if $\sigma \sim_{(\Phi,\phi)} \Lambda$, $\sigma' \sim_{(\Phi,\phi)} \Lambda'$ and $f \in \mathcal{C}^\infty(M)$, then

$$\sigma + \sigma' \sim_{(\Phi,\phi)} \Lambda + \Lambda',$$
$$(f \circ \phi) \sigma \sim_{(\Phi,\phi)} f\Lambda.$$

In this way, projectable sections have a natural $\mathcal{C}^\infty(M)$-module structure. However, we need a more general relationship which involves linearity over $\mathcal{C}^\infty(M')$.

The map $\phi$ determines an algebra morphism

$$\phi^* : \mathcal{C}^\infty(M) \to \mathcal{C}^\infty(M'),$$

given by

$$\phi^*(f) = f \circ \phi, \quad \forall f \in \mathcal{C}^\infty(M).$$

Then, $\phi^*$ provides a structure of $\mathcal{C}^\infty(M)$-module to the space $\mathcal{C}^\infty(M') \times \Gamma(A)$. In this way, we can consider $\mathcal{C}^\infty(M') \otimes \Gamma(A)$, where the tensor product is over $\mathcal{C}^\infty(M)$.

**Lemma 4.0.15.** *Let $\phi^*\pi : \phi^*A \to M'$ be the pullback bundle. Then, $\Gamma(\phi^*A)$ is isomorphic, as a $\mathcal{C}^\infty(M')$-module, to $\mathcal{C}^\infty(M') \otimes \Gamma(A)$. The isomorphism $F : \mathcal{C}^\infty(M') \otimes \Gamma(A) \to \Gamma(\phi^*A)$ is characterized by*

$$f' \otimes \Lambda \mapsto f'\overline{\Lambda},$$

*where $\overline{\Lambda} \in \Gamma(\phi^*A)$ is given by*

$$\overline{\Lambda}(x) = (x, \Lambda(\phi(x))), \quad \forall x \in M'.$$

Thus, with $\Phi^* : \Gamma(A') \to \Gamma(\phi^*A)$, for each $\Lambda' \in \Gamma(A')$, we can write

$$\Phi^*(\Lambda') = \sum_{i=1}^{k} F(f'_i \otimes \Lambda_i), \tag{4.6}$$

for suitable $f'_i \in \mathcal{C}^\infty(M')$ and $\Lambda_i \in \Gamma(A)$, but such a representation does not need to be unique.

If we identify $\Gamma(\phi^*A)$ with the module $\Gamma_\phi(A)$ of smooth maps $f : M' \to A$ such that

$$\pi \circ f = \phi.$$

Then (4.6) becomes

$$\Phi \circ \Lambda' = \sum_{i=1}^{k} f'_i(\Lambda_i \circ \phi). \tag{4.7}$$

We refer to relation (4.7) as a $(\Phi, \phi)$-*decomposition* of $\Lambda'$. Note that the statement $\Lambda'$ is $(\Phi, \phi)$-related to $\Lambda$ is equivalent to

$$\Phi^* (\Lambda') = F (1 \otimes \Lambda).$$

Thus, we are ready to give the definition of Lie algebroid morphism.

**Definition 4.0.16.** Let $(A \to M, \sharp, [\cdot, \cdot])$, $(A' \to M', \sharp', [\cdot, \cdot]')$ be Lie algebroids. A *morphism of Lie algebroids* is a vector bundle morphism $\Phi : A' \to A$, $\phi : M' \to M$ such that

$$\sharp \circ \Phi = T\phi \circ \sharp', \tag{4.8}$$

and such that for arbitrary $\Lambda'_1, \Lambda'_2 \in \Gamma(A')$ with $(\Phi, \phi)$-decompositions

$$\Phi \circ \Lambda'_1 = f_i(\Lambda^1_i \circ \phi), \quad \Phi \circ \Lambda'_2 = g_j(\Lambda^2_j \circ \phi),$$

we have

$$\Phi \circ \left[ \Lambda'_1, \Lambda'_2 \right] = f_i g_j \left( \left[ \Lambda^1_i, \Lambda^2_j \right] \circ \phi \right)$$
$$+ \Lambda'^{\sharp'}_1 (g_j) \left( \Lambda^2_j \circ \phi \right) - \Lambda'^{\sharp'}_2 (f_i) \left( \Lambda^1_i \circ \phi \right). \tag{4.9}$$

In fact, the right-hand side of Eq. (4.9) is independent of the choice of the $(\Phi, \phi)$-decompositions of $\Lambda'_1$ and $\Lambda'_2$.

Now consider two morphisms of Lie algebroids, $\Phi' : A'' \to A'$, $\phi' : M'' \to M'$ and $\Phi : A' \to A$, $\phi : M' \to M$. One can observe that a $(\Phi', \phi')$-decomposition,

$$\Phi' \circ \Lambda''_1 = f''_i(\Lambda'_{1i} \circ \phi'),$$

combines with a $(\Phi, \phi)$-decomposition of each $\Lambda'_{1i}$ to give a $(\Phi \circ \Phi', \phi \circ \phi')$-decomposition, and verifies (4.9) for decompositions so formed. Therefore, checking that the condition for the anchor is satisfied, we have a category of Lie algebroids. We will denote this category by $\mathcal{LA}$.

**Remark 4.0.17.** In particular, if $\Lambda'_1 \sim_{(\Phi, \phi)} \Lambda_1$ and $\Lambda'_2 \sim_{(\Phi, \phi)} \Lambda_2$, then Eq. (4.9) reduces to

$$\Phi \circ \left[ \Lambda'_1, \Lambda'_2 \right] = [\Lambda_1, \Lambda_2] \circ \phi.$$

On the other hand, if $M = M'$ and $\phi = Id_M$ then Eq. (4.9) reduces to

$$\Phi \circ \left[ \Lambda'_1, \Lambda'_2 \right] = \left[ \Phi \circ \Lambda'_1, \Phi \circ \Lambda'_2 \right], \quad \forall \Lambda'_1, \Lambda'_2 \in \Gamma(A').$$

Next, we are going to introduce the notion of Lie subalgebroid.

**Definition 4.0.18.** Let $(A \to M, \sharp, [\cdot, \cdot])$ be a Lie algebroid. Suppose that $A'$ is an embedded submanifold of $A$ and $M'$ is an immersed submanifold of $M$ with inclusion maps $i_{A'} : A' \hookrightarrow A$ and $i_{M'} : M' \hookrightarrow M$. $A'$ is called a *Lie subalgebroid of $A$* if $A'$ is a Lie algebroid on $M'$ which is a vector subbundle of $\pi_{|M'}$, where $\pi : A \to M$ is the projection map of $A$, equipped with a Lie algebroid structure such that the inclusion is a morphism of Lie algebroids. A *reduced subalgebroid of $A$* is a transitive Lie subalgebroid with $M$ as the base manifold.

**Remark 4.0.19.** Suppose that $M' \subseteq M$ is a closed submanifold then, using the $(i_{A'}, i_{M'})$-decomposition and extending functions, we have that for all $\Lambda_1' \in \Gamma(A')$ there exists $\Lambda_1 \in \Gamma(A)$ such that

$$i_{A'} \circ \Lambda_1' = \Lambda_1 \circ i_{M'}.$$

So, Eq. (4.9) reduces to

$$i_{A'} \circ [\Lambda_1', \Lambda_2']_{M'} = [\Lambda_1, \Lambda_2]_M \circ i_{M'}, \quad \forall \Lambda_1', \Lambda_2' \in \Gamma(A').$$

**Example 4.0.20.** Let $(A \to M, \sharp, [\cdot, \cdot])$ be a Lie algebroid over $M$. Then from Lemma 4.0.2 and Remark 4.0.17 we deduce that the anchor map $\sharp : A \to TM$ is a Lie algebroid morphism from $A$ to the tangent algebroid of $M$.

**Example 4.0.21.** Let $(A \to M, \sharp, [\cdot, \cdot])$ be a Lie algebroid over $M$ and $z$ a point of $M$. Then the inclusion map from the isotropy algebra $\mathrm{Ker}\,(\sharp_z)$ of $z$ to $A$ is a Lie algebroid morphism.

**Example 4.0.22.** Let $\phi : M_1 \to M_2$ be a smooth map. Then $(T\phi, \phi)$ is a Lie algebroid morphism between the tangent algebroids $TM_1$ and $TM_2$.

**Example 4.0.23.** Let $\tau : P \to M$ be a principal bundle with structure group $G$ and $(A \to P, \sharp, [\cdot, \cdot])$ be a Lie algebroid (with vector bundle projection $\pi : A \to P$) in the conditions of Example 4.0.12. If $\overline{\pi} : A/G \to M$ is the quotient Lie algebroid by the action of the Lie group $G$ then $(\overline{\tau}, \tau)$ is a Lie algebroid morphism. Remember that $\overline{\tau}$ is the quotient projection $\overline{\tau} : A \to A/G$.

### Construction of the associated Lie algebroid

Now, it is time to justify the name of *infinitesimal groupoid* which was initially given to Lie algebroids. In order to do this, we will generalize the construction of the Lie algebra of a Lie group. As an important case of this construction we find the *1-jets algebroid*.

The process of construction of the associated Lie algebroid to a Lie groupoid was formally extended for the case of subgroupoids (not necessarily Lie subgroupoids) of Lie groupoids in Jiménez *et al.* (2018). This paper is crucial for the development exposed in this book and it will be properly explained in Chapter 6, Section 6.1.

As a first step, we should generalize the notion of left-invariant vector fields of a Lie group.

**Definition 4.0.24.** Let $\Gamma \rightrightarrows M$ be a Lie groupoid with target map $\beta$. A vector field $\Theta \in \mathfrak{X}(\Gamma)$ is called *left-invariant* if it satisfies the following two properties:

(a)  $\Theta$ is tangent to the $\beta$-fibres $\Gamma^x$, for all $x \in M$.
(b)  For each $g \in \Gamma$, the left translation $L_g$ preserves $\Theta$.

Denote the space of smooth left-invariant vector fields on $\Gamma$ by $\mathfrak{X}_L(\Gamma)$.

Similarly to the case of Lie groups, it is clear that the Lie bracket of two left-invariant vector fields is again a left-invariant vector field, say

$$[\mathfrak{X}_L(\Gamma), \mathfrak{X}_L(\Gamma)] \subset \mathfrak{X}_L(\Gamma). \tag{4.10}$$

On the other hand, $T\beta$ has constant rank. Thus, we may define the vector subbundle of $T\Gamma$ given by

$$\bigsqcup_{x \in M} T\Gamma^x = \bigsqcup_{g \in \Gamma} \mathrm{Ker}\,(T_g\beta) = \mathrm{Ker}\,(T\beta).$$

Let $\epsilon : M \to \Gamma$ be the section of identities. We define the pullback vector bundle on $M$,

$$\epsilon^*\,(\mathrm{Ker}\,(T\beta)) = M \times_{\epsilon,\pi_\Gamma} \mathrm{Ker}\,(T\beta), \tag{4.11}$$

where $\pi_\Gamma : T\Gamma \to \Gamma$ is the tangent bundle projection on $\Gamma$ and $M \times_{\epsilon,\pi_\Gamma} \mathrm{Ker}\,(T\beta)$ is the pullback space according to the following diagram

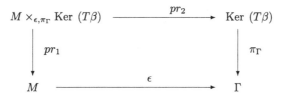

where $pr_i$ is the projection on the $i$-component of $M \times_{\epsilon, \pi_\Gamma} \mathrm{Ker}\,(T\beta)$. Notice that $M \times_{\epsilon, \pi_\Gamma} \mathrm{Ker}\,(T\beta)$ can be depicted as the disjoint union,

$$\bigsqcup_{x \in M} \mathrm{Ker}\,\left(T_{\epsilon(x)}\beta\right). \tag{4.12}$$

We will denote this disjoint union by $A\Gamma$ and the projection will be denoted by $\pi^\epsilon : A\Gamma \to M$. Note that the sections of $A\Gamma$ are determined by smooth maps $\Lambda : M \to T\Gamma$ such that

(i) $T\beta \circ \Lambda = 0$;
(ii) $\pi_\Gamma \circ \Lambda = \epsilon$.

Thus, for each map $\Lambda \in \Gamma\,(A\Gamma)$ we can define the left-invariant vector field on $\Gamma$ given by

$$\Theta^\Lambda\,(g) = T_{\epsilon(\alpha(g))}L_g\,(\Lambda\,(\alpha\,(g))), \quad \forall g \in \Gamma,$$

i.e., $\Theta^\Lambda$ is determined by the following equality

$$\Theta^\Lambda\,(\epsilon\,(x)) = \Lambda\,(x), \quad \forall x \in M.$$

Conversely, if $\Theta \in \mathfrak{X}_L\,(\Gamma)$, then $\Lambda^\Theta = \Theta \circ \epsilon : M \to T\Gamma$ induces a section of $A\Gamma$ and, indeed, the correspondence $\Lambda \mapsto \Theta^\Lambda$ generates a linear isomorphism from $\Gamma\,(A\Gamma)$ to $\mathfrak{X}_L\,(\Gamma)$. With this identification $\Gamma\,(A\Gamma)$ inherits a Lie bracket from $\mathfrak{X}_L\,(\Gamma)$.

This construction is a natural extension of the Lie structure in the associated Lie algebra of a Lie group. In that case, we fix a Lie group $G$ and an element $\xi$ of $T_e G$. Then, we constructed the associated left-invariant vector field by the equality

$$\Theta^\xi\,(e) = \xi.$$

Using this equality $T_e G$ is endowed with a Lie algebra structure.

Finally, an anchor map can be defined as follows: identify $\mathcal{C}^\infty\,(M)$ with the space $\mathcal{C}^\infty_L\,(\Gamma)$ of left-invariant functions on $\Gamma$ using the map given by $\Phi : f \in \mathcal{C}^\infty\,(M) \mapsto f \circ \alpha \in \mathcal{C}^\infty_L\,(\Gamma)$ ($f \circ \alpha \in \mathcal{C}^\infty_L\,(\Gamma)$ because

$\alpha\left(g\cdot h\right)=\alpha\left(h\right)$, for all $\left(g,h\right)\in\Gamma_{(2)})$ with inverse map $\Phi^{-1}:f\in\mathcal{C}_{L}^{\infty}\left(\Gamma\right)\mapsto f\circ\epsilon\in\mathcal{C}^{\infty}\left(M\right)$.

Furthermore, like in the case of Lie groups, $\Theta\in\mathfrak{X}_{L}\left(\Gamma\right)$ if and only if

$$\Theta\left(f\circ L_{g}\right)=\Theta\left(f\right)\circ L_{g},\quad\forall g\in\Gamma,\quad\forall f\in\mathcal{C}^{\infty}(\Gamma).$$

So, if $\Theta\in\mathfrak{X}_{L}\left(\Gamma\right)$ and $f\in\mathcal{C}_{L}^{\infty}\left(\Gamma\right)$, then we have

$$\Theta\left(f\right)\in\mathcal{C}_{L}^{\infty}(\Gamma).$$

In this way, we will define the anchor map as follows: let $\Lambda$ be a section of $\Gamma\left(A\Gamma\right)$; then for each $f\in\mathcal{C}^{\infty}\left(M\right)$ we define

$$\Lambda^{\sharp}\left(f\right)=\Theta^{\Lambda}\left(f\circ\alpha\right)\circ\epsilon.$$

Thus, $\Lambda^{\sharp}\left(f\right)\in\mathcal{C}^{\infty}\left(M\right)$ for all $f\in\mathcal{C}^{\infty}\left(M\right)$. Furthermore, its inherits the Leibniz rule from $\Theta^{\Lambda}$ and so, $\sharp$ is well-defined.

Notice that, for each $x\in M$ and $f\in\mathcal{C}^{\infty}\left(M\right)$

$$\begin{aligned}\left\{\Lambda^{\sharp}\left(f\right)\right\}\left(x\right)&=\left\{\Theta^{\Lambda}\left(f\circ\alpha\right)\right\}\left(\epsilon\left(x\right)\right)\\&=\left\{T_{\epsilon(x)}\alpha\left(\Theta^{\Lambda}\left(\epsilon\left(x\right)\right)\right)\right\}\left(f\right)\\&=\left\{T_{\epsilon(x)}\alpha\left(\Lambda\left(x\right)\right)\right\}\left(f\right),\end{aligned}$$

i.e., it satisfies that

$$\sharp\left(\Lambda\left(x\right)\right)=T_{\epsilon(x)}\alpha\left(\Theta^{\Lambda}\left(\epsilon\left(x\right)\right)\right)=T_{\epsilon(x)}\alpha\left(\Lambda\left(x\right)\right),\qquad(4.13)$$

for all $\Lambda\in\Gamma\left(A\Gamma\right)$ and $x\in M$. Hence,

$$\sharp=\left\{T\alpha\right\}_{|A\Gamma}.\qquad(4.14)$$

Therefore, $\sharp$ is a vector bundle morphism and it satisfies the Leibniz rule. So, $\left(A\Gamma\to M,\sharp,\left[\cdot,\cdot\right]\right)$ is a Lie algebroid, called the *Lie algebroid associated to the Lie groupoid* $\Gamma\rightrightarrows M$, and denoted by $A\Gamma$.

**Remark 4.0.25.** Let $\Gamma\rightrightarrows M$ be a Lie groupoid. For any $x\in M$, the associated Lie algebra to the isotropy Lie group $\Gamma_{x}^{x}$, $A\left(\Gamma_{x}^{x}\right)$ is isomorphic to the isotropy Lie algebra through $x$, i.e.,

$$A\left(\Gamma_{x}^{x}\right)\cong\mathrm{Ker}\left(\sharp_{x}\right).\qquad(4.15)$$

Now, we can prove a result which shows the real nature of the given relation between Lie groupoids and Lie algebroid. This result will be proved for (not necessarily Lie) subgroupoids of a given Lie groupoid in Chapter 6, Section 6.1.

**Theorem 4.0.26.** *There is a natural functor A from the category of Lie groupoids to the category of Lie algebroids.*

**Proof.** We already have given the definition of the correspondence between objects $(\Gamma \rightrightarrows M \to A\Gamma)$ and we will obtain the correspondence between morphisms.

Let $(\Phi, \phi) : \Gamma_1 \rightrightarrows M_1 \to \Gamma_2 \rightrightarrows M_2$ be a Lie groupoid morphism, with $\Phi : \Gamma_1 \to \Gamma_2$ and $\phi_1 : M_1 \to M_2$. Then, $(\Phi, \phi)$ induces a morphism of Lie algebroids from $A\Gamma_1$ to $A\Gamma_2$ given by $(\Phi_*, \phi)$ where

$$\Phi_* = T\Phi_{|A\Gamma_1}. \tag{4.16}$$

So, if $v_{\epsilon_1(x)} \in \mathrm{Ker}\left(T_{\epsilon_1(x)}\beta_1\right)$ then

$$\Phi_*\left(v_{\epsilon_1(x)}\right) = T_{\epsilon_1(x)}\Phi_x(v_{\epsilon_1(x)}),$$

where $\Phi_x : \beta_1^{-1}(x) \to \beta_2^{-1}(\phi(x))$, for each $x \in M$. Now,

$$\pi_{\Gamma_2}\left(T_{\epsilon_1(x)}\Phi\left(v_{\epsilon_1(x)}\right)\right) = \Phi(\epsilon_1(x)),$$

and using that $(\Phi, \phi)$ is a morphism of Lie groupoids, it follows

$$\pi_{\Gamma_2}\left(T_{\epsilon_1(x)}\Phi\left(v_{\epsilon_1(x)}\right)\right) = \epsilon_2(\phi(x)).$$

Furthermore, using again that $(\Phi, \phi)$ is a morphism of Lie groupoids,

$$T_{\Phi(\epsilon_1(x))}\beta_2\left(T_{\epsilon_1(x)}\Phi\left(v_{\epsilon_1(x)}\right)\right) = T_{\epsilon_1(x)}\left(\beta_2 \circ \Phi\right)\left(v_{\epsilon_1(x)}\right)$$
$$= T_{\epsilon_1(x)}\left(\phi \circ \beta_1\right)\left(v_{\epsilon_1(x)}\right) = 0.$$

Thus, $\Phi_*\left(v_{\epsilon_1(x)}\right) \in A\Gamma_2$, i.e.,

$$\Phi_* : A\Gamma_1 \to A\Gamma_2.$$

Also, it is trivial to show that the following diagram is commutative:

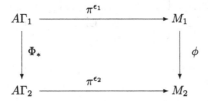

In addition, for all $x \in M_1$ the map

$$\Phi_{*|A\Gamma_{1x}} : A\Gamma_{1x} \to A\Gamma_{2x},$$

is linear so that we have that the map $(\Phi_*, \phi)$ is a vector bundle morphism. Finally, we must study how $(\Phi_*, \phi)$ works with the anchor map and the bracket of sections:

(a)  Observe that, for each $i = 1, 2$ we have

$$\sharp_i (\Lambda (x)) = T_{\epsilon_i(x)} \alpha_i (\Lambda (x)),$$

for any $\Lambda \in \Gamma (A\Gamma_i)$ and $x \in M_i$. Using this identity it is trivial that the following diagram

$$
\begin{array}{ccc}
A\Gamma_1 & \xrightarrow{\ \Phi_* \ } & A\Gamma_2 \\
\downarrow{\scriptstyle \sharp_1} & & \downarrow{\scriptstyle \sharp_2} \\
TM_1 & \xrightarrow{\ T\phi \ } & TM_2
\end{array}
$$

is a commutative diagram, i.e., $\sharp_2 \circ \Phi_* = T\phi \circ \sharp_1$.

(b)  Let $\Lambda$ be a section of $A\Gamma_1$ with $(\Phi_*, \phi)$-decomposition,

$$\Phi_*\Lambda = f_i (\Lambda_i \circ \phi). \tag{4.17}$$

Then, for all $g \in \Gamma_1$,

$$\{T\Phi \circ \Theta^\Lambda\} (g) = T_g \Phi \left( T_{\epsilon_1(\alpha_1(g))} L_g (\Lambda (\alpha_1 (g))) \right)$$

$$= T_{\epsilon_1(\alpha_1(g))} (\Phi \circ L_g) (\Lambda (\alpha_1 (g))).$$

Since $(\Phi, \phi)$ is a morphism of Lie groupoids

$$\Phi \circ L_g = L_{\Phi(g)} \circ \Phi.$$

Then,

$$\{T\Phi \circ \Theta^\Lambda\}(g) = T_{\epsilon_1(\alpha_1(g))}\left(L_{\Phi(g)} \circ \Phi\right)(\Lambda(\alpha_1(g)))$$

$$= T_{\Phi(\epsilon_1(\alpha_1(g)))}L_{\Phi(g)}\left(T_{\epsilon_1(\alpha_1(g))}\Phi(\Lambda(\alpha_1(g)))\right)$$

$$= T_{\Phi(\epsilon_1(\alpha_1(g)))}L_{\Phi(g)}\{(T\Phi \circ \Lambda)(\alpha_1(g))\}$$

$$= (f_i \circ \alpha_1(g))\{T_{\Phi(\epsilon_1(\alpha_1(g)))}L_{\Phi(g)}\}(\Lambda_i \circ \phi \circ \alpha_1(g))$$

$$= (f_i \circ \alpha_1(g))\{T_{\Phi(\epsilon_1(\alpha_1(g)))}L_{\Phi(g)}\}(\Lambda_i \circ \alpha_2 \circ \Phi(g))$$

$$= (f_i \circ \alpha_1(g))(\Theta^{\Lambda_i}(\Phi(g))).$$

Thus, we have got the following identity

$$T\Phi \circ \Theta^\Lambda = (f_i \circ \alpha_1)(\Theta^{\Lambda_i} \circ \Phi).$$

Finally, using this identity and that $(T\Phi, \Phi)$ is a Lie algebroid morphism between the tangent algebroids, it is a routinary exercise to prove the identity (4.9). □

The morphism induced by a morphism $(\Phi, \phi)$ of Lie groupoids over the associated Lie algebroids will be denoted by $A\Phi$.

Now, we are going to give some examples of the above general construction.

**Example 4.0.27.** Let $M$ be a smooth manifold and $M \times M \rightrightarrows M$ be the pair groupoid (see Example 3.0.6). Then, the vector bundle $\epsilon^*(\text{Ker}(t\beta))$ can be seen as the tangent bundle $\pi_M : TM \to M$. Then, it follows that the associated Lie algebroid to $M \times M \rightrightarrows M$ is the tangent algebroid.

**Example 4.0.28.** Let $M$ be a manifold and $G$ be a Lie group. Consider the trivial Lie groupoid on $M$ with group $G$ (see Example 3.0.18). Then, the associated Lie algebroid is the trivial Lie algebroid on $M$ with structure algebra $\mathfrak{g}$ (see Example 4.0.8), i.e., $TM \oplus (M \times \mathfrak{g}) \to M$.

**Example 4.0.29.** Let $\pi : P \to M$ be a principal bundle with structure group $G$. Denote by $\phi : G \times P \to P$ the action of $G$ on $P$.

Now, suppose that $\Gamma \rightrightarrows P$ is a Lie groupoid, with $\overline{\phi} : G \times \Gamma \to \Gamma$ a free and proper action of $G$ on $\Gamma$ such that, for each $g \in G$, the pair $(\overline{\phi}_g, \phi_g)$ is an isomorphism of Lie groupoids. So, we may construct the quotient Lie groupoid by the action of a Lie group, $\Gamma/G \rightrightarrows M$ (see Example 3.0.20).

Then, by construction, we may identify $A(\Gamma/G)$ with the quotient Lie algebroids by the action of a Lie group, $A\Gamma/G$ (see Example 4.0.12).

As a particular case, we may give the following interesting example.

**Example 4.0.30.** Let $\pi : P \to M$ be a principal bundle with structure group $G$ and $\text{Gauge}(P)$ be the Gauge groupoid (see Example 3.0.21). Then, the associated Lie algebroid to $Gauge(P)$ is the Atiyah algebroid associated with the principal bundle $\pi : P \to M$ (see Example 4.0.13).

**Example 4.0.31.** Let $\Phi(A) \rightrightarrows M$ be the frame groupoid. Then $A\Phi(A)$ is called *frame algebroid* (see Example 3.0.22). As a particular case, $A\Pi^1(M, M)$ is called *1-jets algebroid*.

Let $(x^i)$ be a local coordinate system defined on some open subset $U \subseteq M$, using Eqs. (3.3) and (4.11) we can consider local coordinates on $A\Pi^1(M, M)$ as follows:

$$A\Pi^1(U, U) : \left(\left(x^i, x^i, \delta_j^i\right), 0, v^i, v_j^i\right) \cong \left(x^i, v^i, v_j^i\right). \qquad (4.18)$$

We will pay an special attention to the 1-jet algebroid because of the fundamental role which will play in Chapter 5 (see Jiménez *et al.*, 2019).

Let us describe a particular but important family of section of $A\Pi^1(M, M)$. Consider a vector field $\Theta$ on $M$. Denote by $\varphi_t^\Theta : U_t \to U_{-t}$ the (local) flow of $\Theta$. Then, for each $t$ we can construct a diffeomorphism,

$$\Pi\varphi_t^\Theta : \Pi^1(U_{-t}, \mathcal{B}) \to \Pi^1(U_t, \mathcal{B}),$$

such that

$$\Pi\varphi_t^\Theta(g) = g \cdot j^1_{\varphi_{-t}^\Theta(\alpha(g)), \alpha(g)}\varphi_t^\Theta.$$

So, this flow induces a left-invariant vector field on $\Pi^1(M, M)$ which generates a section of $A\Pi^1(M, M)$ denoted by $j^1\Theta$. $j^1\Theta$ is called the *complete lift* of $\Theta$ on $\Pi^1(M, M)$.

Let $\left(x^i\right)$ be a local chart of $M$ and $(x^i, y^j, y_i^j)$ be the induced local chart of $\Pi^1(M, M)$. Assume that, locally, $\Theta$ is written as follows:

$$\Theta = \Theta^i \frac{\partial}{\partial x^i}.$$

Then, locally, $j^1 \Theta$ is expressed in the following way:

$$j^1 \Theta = -\Theta^i \frac{\partial}{\partial x^i} + y_k^j \frac{\partial \Theta^k}{\partial x^i} \frac{\partial}{\partial y_i^j} \tag{4.19}$$

Notice that $j^1 \Theta$ can be equivalently induced by a 1-jet of $\Theta$. Thus, $A\Pi^1(M, M)$ can be interpreted as the bundle of 1-jets of vector fields on $M$.

Let us now give an alternative description for the 1-jets algebroid. The process shown here is explained for the frame algebroid in Kosmann-Schwarzbach and Mackenzie (2002) and Mackenzie (2005).

Let Der $(TM)$ be the collection of all derivations on $M$. Remember that (see Appendix B) a derivation $D : \mathfrak{X}(M) \to \mathfrak{X}(M)$ on $M$ is an $\mathbb{R}$-linear map with base vector field $\Theta \in \mathfrak{X}(M)$ such that for each $f \in \mathcal{C}^\infty(M)$ and $\Xi \in \mathfrak{X}(M)$,

$$D(f\Xi) = fD(\Xi) + \Theta(f)\Xi.$$

- A *zeroth-order differential operator on $M$* is a $\mathcal{C}^\infty(M)$-linear endomorphism $\mathfrak{X}(M) \to \mathfrak{X}(M)$.
- A *first-order differential operator on $M$* is a $\mathbb{R}$-linear map $D : \mathfrak{X}(M) \to \mathfrak{X}(M)$ such that for each $f \in \mathcal{C}^\infty(M)$, the map

$$\mathfrak{X}(M) \to \mathfrak{X}(M)$$

$$\Xi \mapsto D(f\Xi) - fD(\Xi).$$

is a zeroth-order differential operator on $M$. Equivalently, for all $f, g \in \mathcal{C}^\infty(M)$ and $\Xi \in \mathfrak{X}(M)$,

$$D(fg\Xi) = fD(g\Xi) + gD(f\Xi) - fgD(\Xi).$$

Notice that the space of zeroth-order differential operators is contained in the space of derivations on $M$ ($X = 0$) and this space is contained in the space of first-order differential operators.

Now, associated to any first-order differential operator, $D$, there is a map from 1-forms on $M$ to zeroth-order differential operators on $M$, called *symbol of D*, which is determined by

$$\{\sigma(D)(df)\}(\Xi) = [D, f](\Xi) = D(f\Xi) - fD(\Xi),$$

for all $f \in \mathcal{C}^\infty(M)$ and $\Xi \in \mathfrak{X}(M)$.

Thus, $D$ is a derivation on $M$ if and only if there exists a vector field $\Theta$ on $M$ such that for all $\Lambda \in \Omega^1(M)$,

$$\sigma(D)(\Lambda) = \Lambda(\Theta)\, Id_{\mathfrak{X}(TM)}.$$

So, the symbol of $D$ evaluated at any 1-form $\Lambda$ at a point $x \in M$ is a scalar multiple of the identity map of the fibre $T_xM$ over $x$; $\sigma(D)(\Lambda)(x) = \Lambda(x)(\Theta(x))\, Id_{T_xM}$. We have thus obtained that a first-order differential operator is a derivation if and only if it has scalar symbol. Furthermore, it is obvious that $\sigma(D) = 0$ if and only if $D$ is a zeroth-order differential operator.

Now, the space of first-order differential operators on $M$ can be considered as the space of sections of a vector bundle $\mathrm{Diff}^1(M)$ on $M$. So, we can define $\sigma$ as a vector bundle morphism

$$\sigma : \mathrm{Diff}^1(M) \to \mathrm{Hom}(T^*M, \mathrm{End}(TM)),$$

which will be called the *symbol of M*.

It turns out that $\sigma$ is a surjective submersion and its kernel the zeroth-order differential operators. Thus, $\sigma$ induces a short exact sequence of vector bundles over $M$

$$\mathrm{End}(TM) \hookrightarrow \mathrm{Diff}^1(M) \to \mathrm{Hom}(T^*M, \mathrm{End}(TM)).$$

Next, we can define $\mathfrak{D}(TM)$ to be the pullback vector bundle defined by the symbol map and the injective map

$$I : TM \to \mathrm{Hom}(T^*M, \mathrm{End}(TM))$$

$$v_x \mapsto I(v_x),$$

where for each $\Lambda_x \in T_x^*M$ and $w_x \in T_xM$

$$\{I(v_x)(\Lambda_x)\}(w_x) = \Lambda_x(v_x)\, w_x,$$

according to the diagram

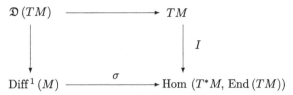

Furthermore, taking into account that the left-hand vertical arrow is an injective immersion we can consider $\mathfrak{D}(TM)$ as a subbundle of $\text{Diff}^1(TM)$. We will denote the top arrow by $a$ and, clearly, as we have noticed before, the kernel of $a$ is $\text{End}(TM)$. So, using $a$, we can consider another exact sequence

$$\text{End}(TM) \hookrightarrow \mathfrak{D}(TM) \to TM,$$

where, taking into account the map $I$, the space of sections of $\mathfrak{D}(TM)$ is, indeed, identifiable with the space $\text{Der}(TM)$ of the derivations on $M$.

In fact, we can endow the vector bundle $\mathfrak{D}(TM)$ with a Lie algebroid structure.

- Let $D_1, D_2$ be derivations on $M$, we can define $[D_1, D_2]$ as the commutator, i.e.,

$$[D_1, D_2] = D_1 \circ D_2 - D_2 \circ D_1.$$

A simple computation shows that the commutator of two derivations is again a derivation, indeed, the base vector field of $[D_1, D_2]$ is given by

$$[\Theta_1, \Theta_2], \tag{4.20}$$

where $\Theta_1$ and $\Theta_2$ are the base vector fields of $D_1$ and $D_2$ respectively.

- Let $D$ be a derivation on $M$, then $D^\sharp$ is its base vector field.

Thus, with this structure $\mathfrak{D}(TM)$ is a transitive Lie algebroid called the *Lie algebroid of derivations on $M$*.

Note that in this Lie algebroid the fibre-wise linear sections of $\sharp$ are $\mathcal{C}^\infty(M)$-linear maps from $\mathfrak{X}(M)$ to $\text{Der}(TM)$. So, the space of fibre-wise linear sections of $\sharp$ is, indeed, the space of covariant derivatives on $M$ (see Appendix B). In fact, it is easy to see that *a covariant derivative* $\nabla$ *is a Lie algebroid morphism (from the tangent algebroid to the algebroid of derivations) if and only if* $\nabla$ *is flat.*

Finally, it is turn to relate this algebroid with the 1-jets Lie algebroid. Consider $\Lambda \in \Gamma \left( A\Pi^1 \left( M, M \right) \right)$ and $\Theta^\Lambda$ its associated left-invariant vector field on $\Pi^1 \left( M, M \right)$. Denote by $\varphi_t^\Lambda : \mathcal{U}_t \to \mathcal{U}_{-t}$ the flow of $\Theta^\Lambda$.

Then, we can define a (local) linear map $\left( \varphi_t^\Lambda \right)^* : \mathfrak{X} \left( M \right) \to \mathfrak{X} \left( M \right)$ satisfying

$$\left\{ \left( \varphi_t^\Lambda \right)^* \left( \Theta \right) \right\} \left( x \right) = \varphi_t^\Lambda \left( \epsilon \left( x \right) \right) \left( \Theta \left( \left( \alpha \circ \varphi_t^\Lambda \right) \left( \epsilon \left( x \right) \right) \right) \right),$$

for each $\Theta \in \mathfrak{X} \left( M \right)$ and $x \in M$. Thus, we can define the following derivation on $M$:

$$D^\Lambda = \frac{\partial}{\partial t_{|0}} \left( \varphi_t^\Lambda \right)^*.$$

In other words, for any $\Theta \in \mathfrak{X} \left( M \right)$ and $x \in M$ we have

$$D^\Lambda \Theta \left( x \right) = \frac{\partial}{\partial t_{|0}} \left( \varphi_t^\Lambda \left( \epsilon \left( x \right) \right) \left( \Theta \left( \left( \alpha \circ \varphi_t^\Lambda \right) \left( \epsilon \left( x \right) \right) \right) \right) \right).$$

Notice that, for all $f \in \mathcal{C}^\infty \left( M \right)$

$$D^\Lambda f\Theta \left( x \right) = \frac{\partial}{\partial t_{|0}} \left( \varphi_t^\Lambda \left( \epsilon \left( x \right) \right) \left( f \left( \left( \alpha \circ \varphi_t^\Lambda \right) \left( \epsilon \left( x \right) \right) \right) \right. \right.$$

$$\left. \Theta \left( \left( \alpha \circ \varphi_t^\Lambda \right) \left( \epsilon \left( x \right) \right) \right) \right) \right)$$

$$= \frac{\partial}{\partial t_{|0}} \left( f \left( \left( \alpha \circ \varphi_t^\Lambda \right) \left( \epsilon \left( x \right) \right) \right) \varphi_t^\Lambda \left( \epsilon \left( x \right) \right) \right.$$

$$\left. \left( \Theta \left( \left( \alpha \circ \varphi_t^\Lambda \right) \left( \epsilon \left( x \right) \right) \right) \right) \right)$$

$$= \Lambda^\sharp \left( x \right) \left( f \right) \Theta \left( x \right) + f \left( x \right) D^\Lambda \Theta \left( x \right).$$

It is immediate to prove that for each $\Theta \in \mathfrak{X} \left( M \right)$ one has that

$$D^{j^1 \Theta} \Xi = \left[ \Theta, \Xi \right], \ \forall \Xi \in \mathfrak{X}(M). \tag{4.21}$$

This construction gives us a linear map between the sections of the 1-jets Lie algebroid and the algebroid of derivations which induces a Lie algebroid isomorphism $\mathcal{D} : A\Pi^1 \left( M, M \right) \to \mathfrak{D} \left( TM \right)$ over the identity map on $M$.

The following result will show how the map $D$ works locally.

**Lemma 4.0.32.** *Let $M$ be a manifold and $\Lambda$ be a section of the 1-jets algebroid with local expression*

$$\Lambda\left(x^i\right) = (x^i, \Lambda^j, \Lambda_i^j).$$

*The matrix $\Lambda_i^j$ is (locally) the associated matrix to $D^\Lambda$, i.e.,*

$$D^\Lambda\left(\frac{\partial}{\partial x^i}\right) = \Lambda_i^j \frac{\partial}{\partial x^j},$$

*and the base vector field of $D^\Lambda$ is $\Lambda^\sharp$ which is given locally by $\left(x^i, \Lambda^j\right)$.*

**Proof.** Let $\Lambda \in \Gamma\left(A\Pi^1\left(M, M\right)\right)$ be a section of the 1-jets algebroid and $\Theta^\Lambda$ its associated left-invariant vector field over $\Pi^1\left(M, M\right)$. Considering the flow of $\Theta^\Lambda$, $\{\varphi_t^\Lambda : \mathcal{U}_t \to \mathcal{U}_{-t}\}$ we have by left invariance that

$$\varphi_t^\Lambda\left(x\right) = \xi^{-1} \cdot \varphi_t^\Lambda\left(\xi\right), \quad \forall x \in \alpha(\mathcal{U}_t),$$

where $\xi \in \mathcal{U}_t \cap \alpha^{-1}\left(x\right)$.

Now, let us take local coordinate systems $\left(x^i\right)$ and $\left(y^j\right)$ and its induced local coordinates over $\Lambda$, then

$$\Lambda\left(x^i\right) = (x^i, \Lambda^j, \Lambda_i^j).$$

Thus, the associated left-invariant vector field is (locally) as follows:

$$\Theta^\Lambda(x^i, y^j, y_i^j) = ((x^i, y^j, y_i^j), \Lambda^j, 0, y_l^j \cdot \Lambda_i^l).$$

Therefore, its flow

$$\varphi_t^\Lambda(x^i, y^j, y_i^j) = (\psi_t^\Lambda(x^i), y^j, y_l^j \cdot \overline{\varphi}_t^\Lambda(x^i)),$$

satisfies that

(i) $\psi_t$ is the flow of $\Lambda^\sharp$;
(ii) $\frac{\partial}{\partial t_{|t=0}}\left(\overline{\varphi}_t^\Lambda\left(x^i\right)\right) = \Lambda_i^j$.

Then,

$$(\varphi_t^\Lambda)^* \left( \frac{\partial}{\partial x^k_{|x^i}} \right) = (\psi_t(x^i), x^i, \overline{\varphi}_t(x^i)) \left( \frac{\partial}{\partial x^k_{|\psi_t(x^i)}} \right)$$

$$= \overline{\varphi}_t(x^i) \frac{\partial}{\partial x^k_{|x^i}}.$$

Hence,

$$D^\Lambda \left( \frac{\partial}{\partial x^k} \right) = \Lambda^j_k \frac{\partial}{\partial x^j},$$

i.e., the matrix $\Lambda^j_i$ is (locally) the associated matrix to $D^\Lambda$.    □

Notice that using this isomorphism, we can consider a one-to-one map from fibre-wise linear sections of $\sharp$ in $A\Pi^1(M, M)$ into covariant derivatives over $M$. Thus, having a fibre-wise linear section $\Delta$ of $\sharp$ in $A\Pi^1(M, M)$ we will denote its associated covariant derivative by $\nabla^\Delta$. Furthermore, $\Delta$ is a morphism of Lie algebroids if and only if $\nabla^\Delta$ is flat.

Let $\Delta$ be a fibre-wise linear section of $\sharp$ in $A\Pi^1(M, M)$ and $\nabla^\Delta$ be its associated covariant derivative. Thus, for each $(x^i)$ local coordinate system on $M$

$$\Delta \left( x^i, \frac{\partial}{\partial x^j} \right) = \left( x^i, \frac{\partial}{\partial x^j}, \Delta^j_i \right),$$

where $\Delta^j_i$ depends on $\frac{\partial}{\partial x^j}$ linearly. Thus, we will change the notation as follows

$$\Delta^j_i \left( x^l, \frac{\partial}{\partial x^k} \right) = \Delta^j_{i,k}(x^l). \tag{4.22}$$

Therefore,

$$\nabla^\Delta_{\frac{\partial}{\partial x^j}} \frac{\partial}{\partial x^i} = D^\Delta \left( \frac{\partial}{\partial x^j} \right) \frac{\partial}{\partial x^i} = \Delta^k_{i,j} \frac{\partial}{\partial x^k},$$

where $\Delta \left( \frac{\partial}{\partial x^j} \right)$ is the (local) section of $A\Pi^1(M, M)$ given by

$$\Delta \left( \frac{\partial}{\partial x^j} \right)(x) = \Delta(x) \left( \frac{\partial}{\partial x^j_{|x}} \right).$$

So, $\Delta^k_{i,j}$ are just the Christoffel symbols of $\nabla^\Delta$.

## *Integrability of Lie algebroids*

Pradines (1967, 1968a,b) proposed the possibility of working on a complete Lie theory for Lie groupoids and Lie algebroids presenting also new results.

As we have mentioned, Pradines generalized the construction of the associated Lie algebra to a Lie group to the case of Lie groupoids in order to introduce the structure of Lie algebroid. In fact, this construction is a functor between these categories (see Theorem 4.0.26). This functor was given by Pradines (1967) and is detailed by Mackenzie (1987) for the case of Lie algebroids with the same base and by Higgins and Mackenzie (1990) for the case of Lie algebroids with different bases.

This Lie functor, in the case of Lie groupoids, preserves several fundamental properties. A natural question is the following: Are the same properties preserved in the case of Lie groupoids and Lie algebroids? It turns out that the answer is negative. In fact, there is a Lie theory for these two kind of objects, presented by Pradines (1966, 1967, 1968a,b), in a collection of notes where the proofs of many results are, in fact, omitted.

In this way, there is a need of extending the three Lie's fundamental theorems (see Duistermaat and Kolk, 2000) for Lie groups and Lie algebras:

**Lie's first fundamental theorem:** Any integrable Lie algebra can be integrated to a simply connected Lie group.

**Lie's second fundamental theorem:** Any morphism between integrable Lie algebras can be integrated to a morphism of Lie groups.

**Lie's third fundamental theorem:** Any Lie algebra can be integrated to a Lie group.

Actually it has been proved that Lie's first fundamental theorem and Lie's second fundamental theorem can be extended to the context of Lie groupoids and Lie algebroids. In order to generalize Lie's third fundamental theorem, Pradines (1968b) presents the next question: Is any Lie algebroid integrable (see Definition 4.0.33)? For a long time people thought that there were not non-integrable Lie algebroids, such as believed in Pradines (1968b). Nevertheless, Almeida and Molino (1985) showed that this assumption did not hold and that there are not integrable Lie algebroids. Crainic and Fernandes (2003) give necessary and sufficient conditions for the integrability of any Lie algebroid.

**Definition 4.0.33.** A Lie algebroid $(A \to M, \sharp, [\cdot, \cdot])$ is called *integrable* if it is isomorphic to the Lie algebroid $A\Gamma$ associated to a Lie groupoid $\Gamma \rightrightarrows M$. If this is the case, then $\Gamma \rightrightarrows M$ is called an integral of $(A \to M, \sharp, [\cdot, \cdot])$.

Note that, if $\mathcal{U} \subseteq \Gamma$ is an open reduced Lie subgroupoid of $\Gamma \rightrightarrows M$, then it is clear that $A\mathcal{U}$ and $A\Gamma$ are isomorphic. From now on, we will assume that $M$ is connected.

**Definition 4.0.34.** A Lie groupoid $\Gamma \rightrightarrows M$ is said to be *target-connected* if $\Gamma^x$ is connected for any $x \in M$. It is said to be *target-simply connected* if each $\Gamma^x$ is connected and simply connected.

**Example 4.0.35.** Let $M$ be a connected smooth manifold and $\mathcal{F}$ be a regular foliation on $M$ (see Appendix A). The *monodromy groupoid* $\mathrm{Mon}\,(M, \mathcal{F})$ is a groupoid over $M$ with the following properties:

(i) For each $x, y \in M$, the set of morphisms from $x$ to $y$ is given by

$$\begin{cases} \Pi_{\mathcal{F}(x)}\,(x, y) & \text{if } y \in \mathcal{F}\,(x), \\ \emptyset & \text{if } y \notin \mathcal{F}\,(x), \end{cases}$$

where $\mathcal{F}\,(x) \in \mathcal{F}$ is the leaf through $x$ and $\Pi_{\mathcal{F}(x)}\,(x, y)$ is the set of homotopy classes (relative to endpoints) of paths in $\mathcal{F}\,(x)$ from $x$ to $y$.

(ii) The multiplication is induced by the concatenation of paths.

In particular, the isotropy groups of the monodromy groupoid are the fundamental groups of the leaves and the orbits are the leaves of $\mathcal{F}$. If $\mathcal{F}$ consists of just one leaf, i.e., the connected manifold $M$ itself, then the groupoid $\mathrm{Mon}\,(M, \mathcal{F})$ is called the *fundamental groupoid* of $M$ which is transitive (provided that $M$ is connected), and its isotropy groups are isomorphic to the fundamental groups of $M$.

Let $\beta : \mathrm{Mon}\,(M, \mathcal{F}) \to M$ be the target map of the monodromy groupoid, then

$$\mathrm{Mon}\,(M, \mathcal{F})^x = \Pi_{\mathcal{F}(x)}\,(x),$$

where $\Pi_{\mathcal{F}(x)}\,(x)$ is the set of path classes of paths in $\mathcal{F}\,(x)$ ending at $x$. Thus, using the proof of the theorem of the existence of the

Universal Covering Space (see, for example, Lee, 2000), we get that $\Pi_{\mathcal{F}(x)}(x)$ is simply connected and that the map $q : \Pi_{\mathcal{F}(x)}(x) \to \mathcal{F}(x)$ given by

$$q([\gamma]) = \gamma(0),$$

is a covering projection. Hence, the monodromy groupoid is a target-simply connected Lie groupoid.

Let $\Gamma \rightrightarrows M$ be a Lie groupoid. Define the set

$$\Gamma^o = \bigsqcup_{x \in M} \Gamma^{ox},$$

where $\Gamma^{ox}$ is the connected component of $\Gamma^x$ with $\epsilon(x) \in \Gamma^{ox}$.

If $g, h \in \Gamma^o$ with $g \in \Gamma^{ox}$ and $h \in \Gamma^{oy}$ then, by connectedness, there exist $\gamma_g : I \to \Gamma^{ox}$ and $\gamma_h : I \to \Gamma^{oy}$ such that $\gamma_g(0) = g, \gamma_g(1) = \epsilon(x), \gamma_h(1) = h$ and $\gamma_h(0) = \epsilon(y)$. Furthermore, using that $M$ is a connected manifold, there exists $\gamma : I \to M$ such that $\gamma(0) = x$ and $\gamma(1) = y$. Thus, the smooth path given by

$$\rho = \gamma_h * (\epsilon \circ \gamma) * \gamma_g : I \to \Gamma^o,$$

being $*$ concatenation, satisfies that $\rho(0) = g$ and $\rho(1) = h$. So, $\Gamma^o$ is a connected subset of $\Gamma$.

On the other hand, considering the vertical distribution for $\beta : \Gamma \to M$, $\mathrm{Ker}(T\beta)$ given by

$$g \mapsto \mathrm{Ker}(T_g\beta), \quad \forall g \in \Gamma,$$

is an integral distribution of rank $k$ whose leaves are the connected components of the $\beta$-fibres of $\Gamma$. Therefore, for each $x \in M$ we may consider a foliation chart of $\epsilon(x)$,

$$\phi : U \to \mathbb{R}^{n-k} \times \mathbb{R}^k.$$

By restricting, we assume that the following fact: If $\Gamma^y \cap U \neq \emptyset$, then $\epsilon(y) \in U$. Hence, by connectedness, it is clear that $U \subseteq \Gamma^o$. Taking $V$ the union of such $U$ we obtain an open neighborhood of $\epsilon(M)$ in $\Gamma$ which is contained in $\Gamma^o$.

Now, $\Gamma^o$ is the union of these leaves of the foliation which intersect the open neighborhood and so is itself open. It follows the following result.

**Lemma 4.0.36.** $\Gamma^o$ *is an open and connected Lie subgroupoid of* $\Gamma \rightrightarrows M$ *over* $M$.

**Proof.**    Taking into account that $\Gamma^o$ is an open subset of $\Gamma$, we only have to verify that the structure maps can be restricted to $\Gamma^o$.    $\square$

Therefore, any Lie groupoid has an open and target-connected Lie subgroupoid and then, the Lie algebroids associated are isomorphic.

**Theorem 4.0.37 (Lie I).** *Let $\Gamma \to M$ be a Lie groupoid. There exist a target-simply connected Lie groupoid $\overline{\Gamma} \rightrightarrows M$ and a morphism of Lie groupoids $\overline{F} : \overline{\Gamma} \to \Gamma$, inducing a Lie algebroid isomorphism $A\overline{\Gamma} \to A\Gamma$.*

**Proof.**    Using Lemma 4.0.36, we can assume that $\Gamma$ is source-connected.

Let $\mathcal{F}$ be the regular foliation of $\Gamma$ given by the $\beta$-fibres, and let Mon $(\Gamma, \mathcal{F})$ be its monodromy groupoid over $\Gamma$ (see Example 4.0.35). Let us denote the target (respectively source) map of Mon $(\Gamma, \mathcal{F})$ by $\beta_M$ (respectively $\alpha_M$). Over the monodromy groupoid can be defined the following equivalence relation

$$[\gamma] \sim [\rho],$$

if and only if there exist $g \in \Gamma$ such that $[L_g \circ \gamma] = [\rho]$. Then, the quotient space defines a Lie groupoid $\overline{\Gamma} = \text{Mon}(\Gamma, \mathcal{F})/\Gamma \rightrightarrows M$. Since any monodromy groupoid is target-simply connected we can check that $\overline{\Gamma} \rightrightarrows M$ is again target-simply connected. Finally, let us consider the smooth map

$$F : \text{Mon}(\Gamma, \mathcal{F}) \to \Gamma,$$

given by $F([\gamma]) = \gamma(1) \cdot \gamma(0)^{-1}$ on arrows of Mon $(\Gamma, \mathcal{F})$. Observe that the restriction of $F$ to the $\beta_M$-fibres of Mon $(\Gamma, \mathcal{F})$ is a covering projection over the $\beta$-fibres of $\Gamma$. Therefore, the restriction of $F$ to the $\beta_M$-fibres of Mon $(\Gamma, \mathcal{F})$ is a local diffeomorphism. Then, for each $g \in \Gamma$ the restriction of $F$ induces an isomorphism

$$T_{\epsilon_M(g)}F_{|\beta_M^{-1}(g)} : \text{Ker}\left(T_{\epsilon_M(g)}\beta_M\right) \to \text{Ker}\left(T_{\epsilon(\beta(g))}\beta\right),$$

where $\epsilon_M$ is the section of identities of Mon $(\Gamma, \mathcal{F})$. This proves that the $\beta_M$-fibres of Mon $(\Gamma, \mathcal{F})$ are isomorphic to the $\beta$-fibres of $\overline{\Gamma}$ such that the factorization map

$$\overline{F} : \overline{\Gamma} \to \Gamma,$$

induces an isomorphism between the $\beta$-fibres of $\overline{\Gamma}$ to the $\beta$-fibres of $\Gamma$. Hence, it is easy to prove that $\left(Id_M, \overline{F}\right)$ is a morphism of Lie groupoids which induces an isomorphism between the associated Lie algebroids. $\qquad\square$

A detailed proof of this result can be found in Moerdijk and Mrčun (2002). The transitive case can be found in Mackenzie (2005). The same methods involved in the construction of the target-simply connected groupoid can be used to prove the following integrability result:

**Proposition 4.0.38.** *Any Lie subalgebroid of an integrable Lie algebroid is integrable.*

Next, we will deal with the Lie's second fundamental theorem which proves that any morphism of integrable Lie algebroids can be integrated to a unique morphism of the integral Lie groupoids, provided that the domain groupoid is source-simply connected. This result has been proved by Mackenzie and Xu (2000) (see also Moerdijk and Mrčun, 2002).

**Theorem 4.0.39 (Lie II).** *Let $\Gamma_1 \rightrightarrows M_1$ and $\Gamma_2 \rightrightarrows M_2$ be Lie groupoids, with $\Gamma_2 \rightrightarrows M_2$ target-simply connected and let $\Phi : A\Gamma_2 \to A\Gamma_1$ be a Lie algebroid morphism over $\phi : M_2 \to M_1$. Then there exists a unique morphism of Lie groupoids $F : \Gamma_2 \to \Gamma_1$ with objects map $\phi$ which integrates $\Phi$.*

**Proof.** Let us consider $P = \Gamma_2 \times_{\phi\circ\alpha_2,\alpha_1} \Gamma_1$ be the pullback of $\alpha_1 : \Gamma_1 \to M_1$ along the map $\phi \circ \alpha_2 : \Gamma_2 \to M_1$. We may consider the projection on the first component $pr_1 : P \to \Gamma_2$. Then, for each $(h_2, h_1) \in P$ it is clear that

$$pr_1^{-1}(pr_1(h_2, h_1)) = \{h_2\} \times \alpha_1^{-1}(\phi(\alpha_2(h_2))).$$

Thus, we may construct the foliation on $P$ given by $\mathcal{G} = pr_1^{-1}(\mathcal{F})$, where $\mathcal{F}$ is the foliation of $\Gamma_2$ given by the $\beta_2$-fibres. It satisfies that $\mathcal{G}$ is a regular foliation such that the dimension of the leaves is $\dim(\Gamma_1) - \dim(M_1) + \dim(\Gamma_2) - \dim(M_2)$ and the tangent spaces at the fibres of $\mathcal{G}$ consist of the vectors $(v_2, v_1) \in T_{h_2}\Gamma_2 \times_{T_{h_2}(\phi\circ\alpha_2), T_{h_1}\alpha_1} T_{h_1}\Gamma_1$ such that $v_2 \in \mathrm{Ker}(T_{h_2}\beta_2)$.

(i) $\{T_{h_2}(\phi \circ \alpha_2)\}(v_2) = T_{h_1}\alpha_1(v_1)$.
(ii) $v_2 \in \mathrm{Ker}(T_{h_2}\beta_2)$.

Notice that, it is clear that the vertical distribution of $pr_1$ is

$$(h_2, h_1) \mapsto \mathrm{Ker}\left(T_{(h_2,h_1)}pr_1\right) = \{0\} \times \mathrm{Ker}\left(T_{h_1}\alpha_1\right).$$

Thus,

$$\mathrm{Ker}\left(T_{(h_2,h_1)}pr_1\right) \subseteq T_{(h_2,h_1)}\mathcal{G}\left(h_2, h_1\right),$$

for all $(h_2, h_1) \in P$ with $\mathcal{G}\left(h_2, h_1\right)$ the leaf of the foliation $\mathcal{G}$ at $(h_2, h_1)$.

On the other hand, for each $(h_2, h_1) \in P$ we may consider the left translations $L_{h_i} : \beta_i^{-1}\left(\alpha_i\left(h_i\right)\right) \to \beta_i^{-1}\left(\beta_i\left(h_i\right)\right)$ for $i = 1, 2$. Then, for all $v \in (A\Gamma_2)_{\alpha_2(h_2)} = T_{\epsilon_2(\alpha_2(h_2))}\beta_2^{-1}\left(\alpha_2\left(h_2\right)\right)$, we can take

$$T_{\epsilon_2(\alpha_2(h_2))}L_{h_2}\left(v\right) \in \mathrm{Ker}\left(T_{h_2}\beta_2\right).$$

Furthermore, using that $\Phi$ is a Lie algebroid morphism, for all $v \in (A\Gamma_2)_{\alpha_2(h_2)}$ it makes sense to take

$$T_{\epsilon_1(\alpha_1(h_1))}L_{h_1}\left(\Phi\left(v\right)\right) \in \mathrm{Ker}\left(T_{h_1}\beta_1\right).$$

Next, define the distribution $\mathfrak{H}$ on $P$, such that the fibres $\mathfrak{H}_{(h_2,h_1)}$ consist of the vectors $\left(T_{\epsilon_2(\alpha_2(h_2))}L_{h_2}\left(v\right), T_{\epsilon_1(\alpha_1(h_1))}L_{h_1}\left(\Phi\left(v\right)\right)\right)$ where $v \in (A\Gamma_2)_{\alpha_2(h_2)}$.

Note that, using that $\Phi$ is a morphism of Lie algebroids over $\phi$ ($\Rightarrow \phi \circ \alpha_2 \circ L_{h_2} = \alpha_1 \circ L_{h_1} \circ \Phi$), $\mathfrak{H}$ is well defined. Hence,

$$\mathfrak{H}_{(h_2,h_1)} \subseteq T_{(h_2,h_1)}\mathcal{G}_{(h_2,h_1)}.$$

Since $L_{h_2}$ is a diffeomorphism, the dimension of $\mathfrak{H}_{(h_2,h_1)}$ is equal to $\dim\left(\Gamma_2\right) - \dim\left(M_2\right)$. Furthermore, we get that for all $(h_2, h_1) \in P$,

$$\mathrm{Ker}\left(T_{(h_2,h_1)}pr_1\right) \cap \mathfrak{H}_{(h_2,h_1)} = \{(0,0)\}.$$

In this way, counting the dimensions, we have

$$T\mathcal{G} = \mathrm{Ker}\left(Tpr_1\right) \oplus \mathfrak{H}.$$

In fact, we may prove that $\mathfrak{H}$ is an integrable distribution. So, we will denote the foliation integrating $\mathfrak{H}$ by $\mathcal{H}$.

Then, for any $x_2 \in M_2$ the restriction of the projection $(pr_1)_{|\mathcal{H}(\epsilon_2(x_2),\epsilon_1(\phi(x_2)))} : \mathcal{H}\left(\epsilon_2\left(x_2\right), \epsilon_1\left(\phi\left(x_2\right)\right)\right) \to \alpha^{-1}\left(x_2\right)$ is a covering projection. In fact, by taking into account that the $\beta_2-$fibres of $\Gamma_2$ are simply connected, the map $(pr_1)_{|\mathcal{H}(\epsilon_2(x_2),\epsilon_1(\phi(x_2)))}$ is a diffeomorphism.

Denote by $\nu_{x_2}$ the inverse of this diffeomorphism. Now the union of the maps $\nu_{x_2}$ gives us a map $\nu : \Gamma_2 \to P$.

Consider the map $F : pr_2 \circ \nu : \Gamma_2 \to \Gamma_1$, where $pr_2$ is the projection on the second component $pr_2 : P \to \Gamma_1$. So, $F$ (together with $\phi$) gives a morphism of Lie groupoids.

Finally, for any $v \in (A\Gamma_2)_{x_2}$ we have

$$T_{\epsilon_2(x_2)}F(v) = T_{\nu_{x_2}(\epsilon_2(x_2))}pr_2(v, \Phi(v)) = \Phi(v).$$

Hence, the induced map of $F$ over the Lie algebroids is $\Phi$, i.e.,

$$AF = \Phi.$$

$\square$

Using this result we can improve the result in Proposition 4.0.38 (see, for instance, Moerdijk and Mrčun, 2006).

**Proposition 4.0.40.** *Any subalgebroid of an integrable algebroid, $A\Gamma$, is integrable by a unique immersed subgroupoid of the groupoid $\Gamma$.*

Let us finish dealing with the third Lie's fundamental theorem.

**Lemma 4.0.41.** *Let $\Gamma \rightrightarrows M$ be a Lie groupoid. Then $\Gamma \rightrightarrows M$ is transitive if and only if the associated Lie algebroid $A\Gamma$ is transitive.*

**Proof.** The key to prove this result is the identity

$$T_{\epsilon(x)}(\mathcal{O}(x)) = \sharp(A\Gamma_x), \tag{4.23}$$

for all $x \in M$, where $A\Gamma_x$ is the fibre through $x$ and $\mathcal{O}(x)$ is the orbit of $x$ (see Definition 3.0.10). Observe that, using that $\alpha$ is an open map, $\mathcal{O}(x)$ is a closed subset of $M$ and hence, if $\mathcal{O}(x)$ is an open of $M$, by connectedness, $\mathcal{O}(x) = M$. $\square$

Let $A$ be an integrable transitive Lie algebroid. Then, using the above result, there exists a transitive Lie groupoid $\Gamma \rightrightarrows M$ such that

$$A \cong A\Gamma.$$

Now, taking into account Corollary 3.0.27, there exists a principal bundle with structural group $G$, $\pi : P \to M$, such that

$$A \cong A(\text{Gauge}(P)) \cong TP/G,$$

where $TP/G$ is the Atiyah algebroid associated with $\pi : P \to M$ (see Example 4.0.13). So, we have proved that any integrable transitive Lie algebroid is isomorphic to an Atiyah algebroid.

Notice that the tangent map to $\pi$, $T\pi : TP \to TM$, induces an epimorphism

$$\overline{T\pi} : TP/G \to TM,$$

given by

$$\overline{T\pi}\left(\tau\left(v\right)\right) = T\pi\left(v\right),$$

for all $v \in TP$, where $\tau : TP \to TP/G$ is the quotient projection (observe that the well definition of $\overline{T\pi}$ is given by the fact that $\pi$ is a principal bundle). Therefore, we have an exact sequence of vector bundles

$$0 \to \left(\mathfrak{g} \times P\right)/G \overset{j}{\to} TP/G \overset{\overline{T\pi}}{\to} TM \to 0,$$

which is just the *Atiyah sequence associated with the principal bundle* $\pi : P \to M$.

**Remark 4.0.42.** Suppose that $\mathfrak{g}$ is the Lie algebra of $G$, $pr_2 : \mathfrak{g} \times P \to P$ is the trivial vector bundle and that the action $\overline{\phi} = (Ad, \phi)$ of $G$ on $\mathfrak{g} \times P$ is given by

$$(Ad, \phi)\left(g, (\xi, p)\right) = \left(Ad_g\left(\xi\right), \phi_g\left(p\right)\right),$$

for all $(g, (\xi, p)) \in G \times (\mathfrak{g} \times P)$ where $Ad : G \times \mathfrak{g} \to \mathfrak{g}$ is the adjoint action of $G$ on $\mathfrak{g}$. Note that the space of sections $\Gamma\left(\mathfrak{g} \times P\right)$ of $\mathfrak{g} \times P$ may be identified with the set of $\pi$-vertical vector fields on $P$. In fact, using that $\pi : P \to M$ is the principal bundle with structural group $G$,

$$\mathrm{Ker}\left(T_p\pi\right) = T_p\left(G \cdot p\right) := \{\xi_P\left(p\right) : \xi \in \mathfrak{g}\},$$

where $\xi_P$ is the infinitesimal generator of $\phi$ associated to $\xi \in \mathfrak{g}$.

So, it is easy to construct the isomorphism between $\Gamma\left(\mathfrak{g} \times P\right)$ and the set of $\pi$-vertical vector fields on $P$. In addition, $\overline{\phi}$ satisfies the conditions (i) and (ii) of Example 4.0.12 and the resultant quotient vector bundle $\overline{pr_1} : (\mathfrak{g} \times P)/G \to M = P/G$ is just *the adjoint bundle associated with the principal bundle* $\pi : P \to M$. Furthermore, if for each $\xi \in \mathfrak{g}$, the map

$$j : (\mathfrak{g} \times P) / G \to TP/G,$$

$$[(\xi, p)] \mapsto [\xi_P (p)],$$

induces a monomorphism between the vector bundles $(\mathfrak{g} \times P) / G$ and $TP/G$.

Thus, $(\mathfrak{g} \times P) / G$ may be considered as a vector subbundle of $TP/G$. In addition, the space $\Gamma ((\mathfrak{g} \times P) / G)$ may be identified with the set of vector fields on $P$ which are vertical and $G$-invariant.

As a particular case, if $\omega \in \Omega^2 (M)$ is a closed 2-form on $M$ consider the transitive Lie algebroid $A = TM \oplus (M \times \mathbb{R}) \to M$ (see Example 4.0.11). Then, the next sequence

$$0 \to M \times \mathbb{R} \overset{i_0}{\to} TM \oplus (M \times \mathbb{R}) \overset{pr_1}{\to} TM \to 0,$$

is a exact sequence, where $i_0 : M \times \mathbb{R} \to TM \oplus (M \times \mathbb{R})$ is defined by

$$i_0 (x, t) = 0 \oplus (x, t), \quad \forall (x, t) \in M \times \mathbb{R}.$$

Now, denoting by

$$\Phi : TM \oplus (M \times \mathbb{R}) \to TP/G$$

the Lie algebroid isomorphism from $A$ to $TP/G$, then the following diagram

$$
\begin{array}{ccc}
TM \oplus (M \times \mathbb{R}) & \overset{pr_1}{\longrightarrow} & TM \\
\Phi \downarrow & & \downarrow Id_{TM} \\
TP/G & \overset{\overline{T\pi}}{\longrightarrow} & TM
\end{array}
$$

is commutative. Hence, using the exact sequences, there exists a smooth map $\varphi : M \times \mathbb{R} \to (\mathfrak{g} \times P) / G$ such that

$$
\begin{CD}
M \times \mathbb{R} @>{i_0}>> TM \oplus (M \times \mathbb{R}) \\
@V{\varphi}VV @VV{\Phi}V \\
(\mathfrak{g} \times P)/G @>{j}>> TP/G
\end{CD}
$$

is a commutative diagram.

On the other hand, consider the map $F : TM \to TP/G$ given by

$$
F(v_x) = \Phi(v_x \oplus (x, 0)), \quad \forall v_x \in T_x M, \ \forall x \in M.
$$

So, identifying $\Gamma(TP/G)$ with the $G$-invariant vector fields, we get a linear map $\cdot^h : \mathfrak{X}(M) \to \mathfrak{X}^G(M)$ and therefore, there exists a connection $\Lambda : TP \to \mathfrak{g}$, such that $\cdot^h$ is the horizontal lifting.

In fact, this connection verifies that

$$
\operatorname{Curv}^\Lambda = \omega. \tag{4.24}
$$

Observe that

$$
A(\Gamma_x^x) \cong \operatorname{Ker}(\natural_x) \cong \mathbb{R}.
$$

So, given that $G = \Gamma_x^x$, $\dim(G) = 1$ and, therefore, we may consider $\operatorname{Curv}^\Lambda$ as a 2-form on $P$ as follows:

$$
\operatorname{Curv}^\Lambda(v_p, w_p) = pr_2\left(\varphi^{-1}\left(\left[\left(\operatorname{Curv}^\Lambda(v_p, w_p), p\right)\right]\right)\right),
$$

for all $v_p, w_p \in T_p P$ and $p \in P$, where $pr_2 : M \times \mathbb{R} \to \mathbb{R}$ is the projection on the second component. Also, notice that $\operatorname{Curv}^\Lambda$ can be seen as a 2-form on $M$. In fact, using that $\operatorname{Curv}^\Lambda(v_p, w_p) = 0$, if $v_p$ (or $w_p$) is vertical. Then, we may define $\operatorname{Curv}^\Lambda : TM \times_{\pi_M, \pi_M} TM \to \mathfrak{g}$ as follows:

$$
\operatorname{Curv}^\Lambda\left(v_{\pi(x)}, w_{\pi(x)}\right) = \operatorname{Curv}^\Lambda(\overline{v}_x, \overline{w}_x),
$$

where $T_x\pi(\overline{v}_x) = v_{\pi(x)}$ and $T_x\pi(\overline{w}_x) = w_{\pi(x)}$. If $T_x\pi(\overline{v}_x) = T_x\pi(\overline{u}_x)$, $\overline{v}_x - \overline{u}_x \in \operatorname{Ker}(T_x\pi)$, i.e., $(\overline{v}_x - \overline{u}_x)$ is vertical. Then,

$$
\operatorname{Curv}^\Lambda(\overline{v}_x - \overline{u}_x, \overline{w}_x) = 0,
$$

and therefore, $\operatorname{Curv}^\Lambda$ is well defined. In this way, one can verify identity (4.24).

Now, we are going to use the classical Weil lemma (see Mackenzie, 2005, Theorem 8.1.3).

**Theorem 4.0.43.** *A closed 2-form* $\omega \in \Omega^2(M)$ *is the curvature of a connection in an* $S^1$*-bundle if and only if*

$$\int_\gamma \omega = \int_{S^1} \gamma^* \omega \in \mathbb{Z}, \quad \forall \gamma \in \mathcal{C}^\infty(S^1, M). \tag{4.25}$$

In this way, to found a counterexample for the third Lie's fundamental theorem for Lie algebroids we only have to take $\omega$ such that (4.25) it is not satisfied. For example, we may take $M = S^2$ together the volume standard form, i.e.,

$$i^* \omega,$$

with $i : S^2 \to \mathbb{R}^3$ the inclusion map and

$$\omega = dx \wedge dy \wedge dz.$$

So, we may found a number $k$ such that

$$\int k i^* \omega,$$

is not an integer number.

The second part of the book is devoted to the use of groupoids and distributions to the study of the constitutive theory of materials.

Let us start with an elastic simple material $\mathcal{B}$ with reference configuration $\phi_0$. As we know, $\mathcal{B}$ has associated a mechanical response $W : \mathcal{B} \times Gl(3, \mathbb{R}) \to V$. Equation (2.4) shows us that $W$ can be defined on the space of (local) configurations in such a way that for each configuration $\phi$ we define

$$W\left(j_{X,x}^1 \phi\right) = W(X, F),$$

where $F$ is the associated matrix to the 1-jet at $\phi_0(X)$ of $\phi \circ \phi_0^{-1}$. In fact, composing $\phi_0$ by the left, we obtain that $W$ may be equivalently described as a differentiable map $W : \Pi^1(\mathcal{B}, \mathcal{B}) \to V$ from the groupoid of 1-jets $\Pi^1(\mathcal{B}, \mathcal{B})$ (see Example 3.0.9) to the vector space $V$ which does not depend on the image point of the 1-jets of $\Pi^1(\mathcal{B}, \mathcal{B})$, i.e., for all $X, Y, Z \in \mathcal{B}$

$$W\left(j_{X,Y}^1 \phi\right) = W\left(j_{X,Z}^1\left(\phi_0^{-1} \circ \tau_{Z-Y} \circ \phi_0 \circ \phi\right)\right), \tag{4.26}$$

for all $j^1_{X,Y}\phi \in \Pi^1(\mathcal{B},\mathcal{B})$, where $\tau_v$ is the translation map on $\mathbb{R}^3$ by the vector $v$. Notice that, using Eq. (4.26), we may define $W$ over $\Pi^1(\mathcal{B},\mathbb{R}^3)$, which could be seen as an open subset of $\Pi^1(\mathbb{R}^3,\mathbb{R}^3)$ given by the 1-jets of local diffeomorphisms from points of $\mathcal{B}$ to points of $\mathbb{R}^3$.

Then, condition of being materially isomorphic is rewritten as follows: *Two material particles $X$ and $Y$ are materially isomorphic if and only if there exists a local diffeomorphism $\psi$ from an open neighborhood $\mathcal{U} \subseteq \mathcal{B}$ of $X$ to an open neighborhood $\mathcal{V} \subseteq \mathcal{B}$ of $Y$ such that $\psi(X) = Y$ and*

$$W(j^1_{Y,\kappa(Y)}\kappa \cdot j^1_{X,Y}\psi) = W(j^1_{Y,\kappa(Y)}\kappa), \qquad (4.27)$$

*for all $j^1_{Y,\kappa(Y)}\kappa \in \Pi^1(\mathcal{B},\mathcal{B})$.*

For any two points $X, Y \in \mathcal{B}$, we will denote by $G(X,Y)$ the collection of all 1-jets $j^1_{X,Y}\psi$ which satisfy Eq. (3.0.16), i.e., $G(X,Y)$ is the family of material isomorphisms for $X$ to $Y$. Remember that, in Chapter 2, we proved that the relation of being "*materially isomorphic*" is an equivalence relation. Indeed, what we proved is that the set $\Omega(\mathcal{B}) = \cup_{X,Y \in \mathcal{B}} G(X,Y)$ may be considered as a groupoid over $\mathcal{B}$ with the composition of 1-jets as composition law. Thus, *without any kind of assumption on the uniformity of $\mathcal{B}$, there exists a unique and canonically-defined groupoid, $\Omega(\mathcal{B})$, encoding all the information about the internal constitution of the material body $\mathcal{B}$.*

In fact, $\Omega(\mathcal{B})$ is a subgroupoid of the 1-jets groupoid $\Pi^1(\mathcal{B},\mathcal{B})$ and it is called the *material groupoid* of $\mathcal{B}$.

Notice that, the material symmetry group $G(X)$ at a body point $X \in \mathcal{B}$ is just the isotropy group of $\Omega(\mathcal{B})$ at $X$. For each $X \in \mathcal{B}$, we will denote the set of material isomorphisms from $X$ to any other point (respectively, from any point to $X$) by $\Omega_X(\mathcal{B})$ (respectively, $\Omega^X(\mathcal{B})$). Finally, we will denote the structure maps of $\Omega(\mathcal{B})$ by $\bar{\alpha}$, $\bar{\beta}$, $\bar{\epsilon}$ and $\bar{i}$ which are just the restrictions of the corresponding ones on $\Pi^1(\mathcal{B},\mathcal{B})$.

As a consequence of the continuity of $W$ we have that, for all $X \in \mathcal{B}$, $G(X)$ is a closed subgroup of $\Pi^1(\mathcal{B},\mathcal{B})^X_X$. Hence, the following result is immediate.

**Proposition 4.0.44.** *Let $\mathcal{B}$ be a simple body. Then, for all $X \in \mathcal{B}$ the symmetry group $G(X)$ is a Lie subgroup of $\Pi^1(\mathcal{B},\mathcal{B})^X_X$.*

This could make us think that $\Omega(\mathcal{B})$ is a Lie subgroupoid of $\Pi^1(\mathcal{B}, \mathcal{B})$. However, this is not true (for instance, the dimensions of the groups of material symmetries could change).

Now, the following result is obvious.

**Proposition 4.0.45.** *Let $\mathcal{B}$ be a body. $\mathcal{B}$ is uniform if and only if $\Omega(\mathcal{B})$ is a transitive subgroupoid of $\Pi^1(\mathcal{B}, \mathcal{B})$.*

Next, by composing appropriately with the reference configuration, smooth uniformity (Definition 2.0.6) may be characterized in the following way.

**Proposition 4.0.46.** *A body $\mathcal{B}$ is smoothly uniform if and only if for each point $X \in \mathcal{B}$ there is an neighborhood $\mathcal{U}$ around $X$ such that for all $Y \in \mathcal{U}$ and $j^1_{Y,X}\phi \in \Omega(\mathcal{B})$ there exists a local section $\mathcal{P}$ of*

$$\overline{\alpha}_X : \Omega^X(\mathcal{B}) \to \mathcal{B},$$

*from $\epsilon(X)$ to $j^1_{Y,X}\phi$.*

For obvious reasons, (local) sections of $\overline{\alpha}_X$ will be called *left fields of material isomorphism at $X$*. On the other hand, local sections of

$$\overline{\beta}^X : \Omega_X(\mathcal{B}) \to \mathcal{B}$$

will be called *right fields of material isomorphism at $X$*. Thus, left (respectively right) fields of material isomorphisms in the sense of Chapter 2 are in a bijective correspondence with these left (respectively right) fields of material isomorphisms via composition with the reference configuration $\phi_0$.

Therefore, $\mathcal{B}$ is smoothly uniform if and only if for each two points $X, Y \in \mathcal{B}$ there are two open subsets $\mathcal{U}, \mathcal{V} \subseteq \mathcal{B}$ around $X$ and $Y$ respectively and $\mathcal{P} : \mathcal{U} \times \mathcal{V} \to \Omega(\mathcal{B}) \subseteq \Pi^1(\mathcal{B}, \mathcal{B})$, a differentiable section of the anchor map $(\overline{\alpha}, \overline{\beta})$. When $X = Y$ it is easy to realize that we can assume $\mathcal{U} = \mathcal{V}$ and $\mathcal{P}$ is a morphism of groupoids over the identity map, i.e.,

$$\mathcal{P}(Z, T) = \mathcal{P}(R, T)\mathcal{P}(Z, R), \quad \forall T, R, Z \in \mathcal{U}.$$

So, we may prove a corollary of Proposition 4.0.44.

**Corollary 4.0.47.** *Let $\mathcal{B}$ be a body. $\mathcal{B}$ is smoothly uniform if and only if $\Omega(\mathcal{B})$ is a transitive Lie subgroupoid of $\Pi^1(\mathcal{B}, \mathcal{B})$.*

**Proof.**  Suppose that $\mathcal{B}$ is smoothly uniform. Fix $j_{X,Y}^1 \psi \in \Omega(\mathcal{B})$ and consider $\mathcal{P} : \mathcal{U} \times \mathcal{V} \to \Omega(\mathcal{B})$, a differentiable section of the anchor map $(\overline{\alpha}, \overline{\beta})$ with $X \in \mathcal{U}$ and $Y \in \mathcal{V}$. Then, we may construct the following bijection

$$\Psi_{\mathcal{U},\mathcal{V}} : \Omega(\mathcal{U}, \mathcal{V}) \to \mathcal{B} \times \mathcal{B} \times G(X,Y)$$

$$j_{Z,T}^1 \phi \mapsto (Z, T, \mathcal{P}(Z,Y) \left[j_{Z,T}^1 \phi\right]^{-1} \mathcal{P}(X,T)),$$

where $\Omega(\mathcal{U}, \mathcal{V})$ is the set of material isomorphisms from $\mathcal{U}$ to $\mathcal{V}$. By using Proposition 4.0.44, we deduce that $G(X,Y)$ is a differentiable manifold. Thus, we can endow $\Omega(\mathcal{B})$ with a differentiable structure of a manifold. Now, the result follows (the converse has been proved in Mackenzie (2005)).  $\square$

This result clarify even more the difference between smooth uniformity and ordinary uniformity. Furthermore, it works as an intuition about the lost of differentiability which could have the material groupoid. In particular, as we have previously said, the material groupoid is not necessarily a Lie subgroupoid of $\Pi^1(\mathcal{B}, \mathcal{B})$ (see examples in Chapter 6). This is a really *important fact* in this book. Indeed, this is the reason because Part II is also divided in *two* chapters. First part (Chapter 5) is based on the assumption of the material groupoid is a Lie subgroupoid of $\Pi^1(\mathcal{B}, \mathcal{B})$. Then, we can use its associated Lie algebroid to prove novel results associated (above all) the homogeneity of the material (Jiménez *et al.*, 2018, 2019). On the other hand, the second part (Chapter 6) is focused on finding new structures to deal with the material groupoid without imposing any condition of differentiability. Thus, it arises the notion of *material distributions* (Epstein *et al.*, 2019; Jiménez *et al.*, 2017, 2020) which are generalized to context of general groupoids (Jiménez *et al.*, 2018).

# Part II
# Material Groupoid

# Chapter 5

# Material Algebroid

As we have commented before, this chapter is mainly based on a kind of assumption over the differentiability of the family of material isomorphisms. In particular, the material groupoid will have the structure of Lie groupoid. The crucial point about the development of this chapter is the associated Lie algebroid, which is the infinitesimal version of the mentioned Lie groupoid. In fact, since the material groupoid encodes the mechanical information of the material body, its homogeneity can be characterized through the properties of its associated Lie algebroid. This is indeed accomplished, and related with the earlier approach developed in Elżanowski *et al.* (1990) and Epstein and de León (1998) in the framework of $G$-structures.

The content of this chapter is based mainly in Jiménez *et al.* (2019). However, there are also novel results we have not been published yet. In particular, the content of the subsection entitled *Homogeneity with G-structures*. There, a series of results are presented comparing the frame bundle of a (arbitrary) manifold $M$ with the 1-jets groupoid of $M$.

## 5.1 Integrability

As a first step we will introduce the notion of integrability of reduced subgroupoids of the 1-jets groupoid which is going to be closely related with the notion of integrability of $G$-structures (see Appendix C).

Note that there exists a Lie groupoids isomorphism $L$ : $\Pi^1(\mathbb{R}^n, \mathbb{R}^n) \to \mathbb{R}^n \times \mathbb{R}^n \times Gl(n, \mathbb{R})$ over the identity map defined by

$$L\left(j^1_{x,y}\phi\right) = \left(x, y, J\phi_{|x}\right), \quad \forall j^1_{x,y}\phi \in \Pi^1\left(\mathbb{R}^n, \mathbb{R}^n\right),$$

where $J\phi_{|x}$ is the Jacobian matrix of $\phi$ at $x$. Another way of expressing this isomorphism is identifying $Gl(n, \mathbb{R})$ with the fibre of $F\mathbb{R}^n$ at 0. Then, the isomorphism is given by

$$L\left(j^1_{x,y}\phi\right) = \left(x, y, j^1_{0,0}\left(\tau_{-y} \circ \phi \circ \tau_x\right)\right),$$

for all $j^1_{x,y}\phi \in \Pi^1\left(\mathbb{R}^n, \mathbb{R}^n\right)$, where $\tau_z$ denote the translation on $\mathbb{R}^n$ by the vector $z \in \mathbb{R}^n$. So, the inverse map satisfies

$$L^{-1}\left(x, y, j^1_{0,0}\Phi\right) = j^1_{x,y}\left(\tau_y \circ \Phi \circ \tau_{-x}\right),$$

for all $j^1_{0,0}\Phi \in Gl(n, \mathbb{R})$. Observe that we are canonically identifying any regular matrix with a unique 1-jet of a local diffeomorphism from 0 to 0.

We have thus obtained a Lie groupoid isomorphism $\Pi^1\left(\mathbb{R}^n, \mathbb{R}^n\right) \cong \mathbb{R}^n \times \mathbb{R}^n \times Gl(n, \mathbb{R})$ over the identity map on $\mathbb{R}^n$. Then, if $G$ is a Lie subgroup of $Gl(n, \mathbb{R})$, we can transport $\mathbb{R}^n \times \mathbb{R}^n \times G$ by this isomorphism to obtain a reduced Lie subgroupoid of $\Pi^1\left(\mathbb{R}^n, \mathbb{R}^n\right)$. This kind of reduced subgroupoid will be called *standard flat* on $\Pi^1\left(\mathbb{R}^n, \mathbb{R}^n\right)$.

Let $U, V \subseteq M$ be two open subsets of $M$. We denote by $\Pi^1\left(U, V\right)$ the open subset of $\Pi^1\left(M, M\right)$ defined by $(\alpha, \beta)^{-1}\left(U \times V\right)$. Note that if $U = V$, then, $\Pi^1\left(U, U\right)$ is in fact the 1-jets groupoid of $U$ and, in this way, our notation is consistent. Furthermore, we are going to think about $\Pi^1\left(U, V\right)$ as the restriction of the Lie groupoid $\Pi^1\left(M, M\right)$ equipped with the restriction of the structure maps (this could not be a Lie groupoid). We will also use this notation for subgroupoids of $\Pi^1\left(M, M\right)$.

**Definition 5.1.1.** A reduced subgroupoid $\Pi^1_G\left(M, M\right)$ of $\Pi^1\left(M, M\right)$ will be called *integrable* if it is locally diffeomorphic to the groupoid $\mathbb{R}^n \times \mathbb{R}^n \times G \rightrightarrows \mathbb{R}^n$, for some Lie subgroup $G$ of $Gl(n, \mathbb{R})$.

Before continuing, we need to explain what we understand by "locally diffeomorphic" in this case. So, $\Pi^1_G\left(M, M\right)$ is locally diffeomorphic to $\mathbb{R}^n \times \mathbb{R}^n \times G \rightrightarrows \mathbb{R}^n$ if for all $x, y \in M$ there exist two open sets $U, V \subseteq M$ with $x \in U$, $y \in V$ and two local charts, $\psi_U : U \to \overline{U}$ and $\psi_V : V \to \overline{V}$, which induce a diffeomorphism

$$\Psi_{U,V} : \Pi_G^1(U, V) \to \overline{U} \times \overline{V} \times G, \qquad (5.1)$$

such that $\Psi_{U,V} = \left( \psi_U \circ \alpha, \psi_V \circ \beta, \overline{\Psi}_{U,V} \right)$, where

$$\overline{\Psi}_{U,V} \left( j_{x,y}^1 \phi \right) = j_{0,0}^1 \left( \tau_{-\psi_V(y)} \circ \psi_V \circ \phi \circ \psi_U^{-1} \circ \tau_{\psi_U(x)} \right),$$

for all $j_{x,y}^1 \phi \in \Pi^1(U, V)$. Notice that, $\Pi_G^1(U, V)$ and $\overline{U} \times \overline{V} \times G$ are Lie groupoids if and only if $U = V$ and $\overline{U} = \overline{V}$. Suppose that $U = V$ and $\overline{U} = \overline{V}$, then, for all $x \in U$, $\Psi_{U,U} \left( j_{x,x}^1 Id \right) \in G$. However, $\Psi_{U,U} \left( j_{x,x}^1 Id \right)$ is not necessarily the identity map and, hence, $\Psi_{U,U}$ is not an isomorphism of Lie groupoids.

It is important to note that, if $\Pi_G^1(M, M)$ were integrable, then the Lie group $G$ would be isomorphic (as Lie group) to any isotropy group of $\Pi_G^1(M, M)$. So, in order to evaluate the integrability of a Lie groupoid $\Pi_G^1(M, M)$ we should choose the isotropy groups as candidates.

**Proposition 5.1.2.** *Let $\Pi_G^1(M, M)$ be a reduced Lie subgroupoid of $\Pi^1(M, M)$. $\Pi_G(M, M)$ is integrable if and only if we can cover $M$ by local charts $(\psi_U, U)$ which induce Lie groupoid isomorphisms from $\Pi_G^1(U, U)$ to the restrictions of the standard flat over $G$ to $\psi_U(U)$.*

**Proof.** On the one hand, suppose that $\Pi^1(M, M)$ is integrable. Let $x_0 \in M$ be a point in $M$ and $\psi_U : U \to \overline{U}$ and $\psi_V : V \to \overline{V}$ be local charts through $x_0$ which induced diffeomorphism is

$$\Psi_{U,V} : \Pi_G^1(U, V) \to \overline{U} \times \overline{V} \times G.$$

For each $y \in U \cap V$,

$$\overline{\Psi}_{U,V} \left( j_{y,y}^1 Id \right) = j_{0,0}^1 \left( \tau_{-\psi_V(y)} \circ \psi_V \circ \psi_U^{-1} \circ \tau_{\psi_U(y)} \right) \in G.$$

Then, for all $j_{x,y}^1 \phi \in \Pi_G^1(U \cap V, U \cap V)$, we have

$$j_{0,0}^1 \left( \tau_{-\psi_U(y)} \circ \psi_U \circ \phi \circ \psi_U^{-1} \circ \tau_{\psi_U(x)} \right)$$

$$= j_{0,0}^1 \left( \tau_{-\psi_U(y)} \circ \psi_U \circ \psi_V^{-1} \circ \tau_{\psi_V(y)} \right)$$

$$\cdot j_{0,0}^1 \left( \tau_{-\psi_V(y)} \circ \psi_V \circ \phi \circ \psi_U^{-1} \circ \tau_{\psi_U(x)} \right) \in G.$$

Therefore, denoting $U \cap V$ by $W$, the map

$$\Psi_{W,W} : \Pi_G^1(W, W) \to \overline{W} \times \overline{W} \times G,$$

is, indeed, a Lie groupoid isomorphism over $\psi_W$ where $\Psi_{W,W} = \left( \psi_W \circ \alpha, \psi_W \circ \beta, \overline{\Psi}_{W,W} \right)$, $\psi_W$ is the restriction of $\psi_U$ to $W$ and for all $j_{x,y}^1 \phi \in \Pi^1 (W, W)$,

$$\overline{\Psi}_{W,W} \left( j_{x,y}^1 \phi \right) = j_{0,0}^1 \left( \tau_{-\psi_W(y)} \circ \psi_W \circ \phi \circ \psi_W^{-1} \circ \tau_{\psi_W(y)} \right).$$

On the other hand, suppose that for each $x \in M$ there exists a local chart $(\psi_U, U)$ through $x$ which induces a Lie groupoid isomorphism over $\psi_U$, namely

$$\Psi_{U,U} : \Pi_G^1 (U, U) \to \overline{U} \times \overline{U} \times G, \tag{5.2}$$

such that $\Psi_{U,U} = \left( \psi_U \circ \alpha, \psi_U \circ \beta, \overline{\Psi}_{U,U} \right)$, where for each $j_{x,y}^1 \phi \in \Pi^1 (U, U)$,

$$\overline{\Psi}_{U,U} \left( j_{x,y}^1 \phi \right) = j_{0,0}^1 \left( \tau_{-\psi_U(y)} \circ \psi_U \circ \phi \circ \psi_U^{-1} \circ \tau_{\psi_U(x)} \right).$$

Take open sets $U, V \subseteq M$ such that there exist $\psi_U$ and $\psi_V$ satisfy Eq. (5.2). Suppose that $U \cap V \neq \emptyset$. Then, for all $x, y \in U \cap V$, we have

$$j_{0,0}^1 \left( \tau_{-\psi_U(y)} \circ \psi_U \circ \psi_V^{-1} \circ \tau_{\psi_V(y)} \right)$$

$$\cdot j_{0,0}^1 \left( \tau_{-\psi_V(x)} \circ \psi_V \circ \psi_U^{-1} \circ \tau_{\psi_U(x)} \right) \in G. \tag{5.3}$$

Fixing $z \in U \cap V$, we consider

$$j_{0,0}^{1} \left( \tau_{-\psi_U(z)} \circ \psi_U \circ \psi_V^{-1} \circ \tau_{\psi_V(z)} \right) = A \in Gl(n, \mathbb{R}).$$

So, we define the diffeomorphism $\overline{\psi}_V = A \cdot \psi_V : V \to A \cdot \overline{V}$. Then, using Eq. (5.3) for all $y \in U \cap V$, we deduce that

$$j_{0,0}^{1} \left( \tau_{-\overline{\psi}_V(y)} \circ \overline{\psi}_V \circ \psi_U^{-1} \circ \tau_{\psi_U(y)} \right)$$

$$= j_{0,0}^1 A \cdot \left( \tau_{-\psi_V(y)} \circ \psi_V \circ \psi_U^{-1} \circ \tau_{\psi_U(y)} \right)$$

$$= A \cdot j_{0,0}^1 \left( \tau_{-\psi_V(y)} \circ \psi_V \circ \psi_U^{-1} \circ \tau_{\psi_U(y)} \right) \in G. \tag{5.4}$$

In this way, we consider

$$\Psi_{U,V} : \Pi_G^1 (U, V) \to \overline{U} \times A \cdot \overline{V} \times G$$

$$j_{x,y}^1 \phi \mapsto \left( \psi_U (x), \overline{\psi}_V (y), \overline{\Psi}_{U,V} \left( j_{x,y}^1 \phi \right) \right).$$

where

$$\overline{\Psi}_{U,V} \left( j_{x,y}^1 \phi \right) = j_{0,0}^1 (\tau_{-\overline{\psi}_V(y)} \circ \overline{\psi}_V \circ \phi \circ \psi_U^{-1} \circ \tau_{\psi_U(x)}).$$

We will check that $\overline{\Psi}_{U,V}$ is well-defined. We fix $j_{x,y}^1 \phi \in \Pi_G^1 (U, V)$. Then, we can consider two cases:

(i) $y \in U \cap V$. Then, using Eq. (5.4)

$$j_{0,0}^1(\tau_{-\overline{\psi}_V(y)} \circ \overline{\psi}_V \circ \phi \circ \psi_U^{-1} \circ \tau_{\psi_U(x)})$$

$$= j_{0,0}^1(\tau_{-\overline{\psi}_V(y)} \circ \overline{\psi}_V \circ \psi_U^{-1} \circ \tau_{\psi_U(y)})$$

$$\cdot j_{0,0}^1\left(\tau_{-\psi_U(y)} \circ \psi_U \circ \phi \circ \psi_U^{-1} \circ \tau_{\psi_U(x)}\right) \in G.$$

(ii) $y \notin U \cap V$. Then,

$$j_{z,x}^1\left(\psi_V^{-1} \circ \tau_{\psi_V(z) - \psi_V(y)} \circ \psi_V \circ \phi\right) = j_{z,x}^1 \phi_z,$$

which is in $\Pi_G^1(M, M)$. Hence,

$$j_{0,0}^1(\tau_{-\overline{\psi}_V(y)} \circ \overline{\psi}_V \circ \phi \circ \psi_U^{-1} \circ \tau_{\psi_U(x)})$$

$$= A \cdot j_{0,0}^1\left(\tau_{-\psi_V(y)} \circ \psi_V \circ \phi \circ \psi_U^{-1} \circ \tau_{\psi_U(x)}\right)$$

$$= A \cdot j_{0,0}^1\left(\tau_{-\psi_V(z)} \circ \psi_V \circ \phi_z \circ \psi_U^{-1} \circ \tau_{\psi_U(x)}\right)$$

$$= j_{0,0}^1(\tau_{-\overline{\psi}_V(z)} \circ \overline{\psi}_V \circ \phi_z \circ \psi_U^{-1} \circ \tau_{\psi_U(x)}) \in G.$$

Thus, it is immediate to prove that $\Psi_{U,V}$ is a diffeomorphism which commutes with the restrictions of the structure maps.

Finally, if $U \cap V = \emptyset$ we can find a finite family of local neighborhoods $\{V_i\}_{i=1,\ldots,k}$ such that

(i) $U = V_1$
(ii) $V = V_k$
(iii) $V_i \cap V_{i+1} \neq \emptyset$, $\forall i$

Thus, we can find $\Psi_{U,V}$ following a similar procedure to the one used above. $\qquad \square$

Notice that the difference between the definition of integrability and the equivalent property proved in this proposition is subtle but important. In fact, roughly speaking, we are asserting that we may choose $U$ and $V$ (in Definition 5.1.1 and Eq. (5.1)) such that $U = V$. However, both versions of the definition of integrability will be used in what follows.

**Remark 5.1.3.** Let $\Pi_G^1(M, M)$ be an integrable subgroupoid of $\Pi^1(M, M)$ by the Lie subgroup $G$ of $Gl(n, \mathbb{R})$, i.e., locally diffeomorphic to $\mathbb{R}^n \times \mathbb{R}^n \times G$. Suppose that there exists another subgroup $\tilde{G}$ of $Gl(n, \mathbb{R})$ such that $\Pi_G^1(M, M)$ is locally diffeomorphic to

$\mathbb{R}^n \times \mathbb{R}^n \times \tilde{G}$. Then, using the above result, it is easy to see that $G$ and $\tilde{G}$ are conjugated subgroups of $Gl(n, \mathbb{R})$. Conversely, if $G$ and $\tilde{G}$ are conjugated subgroups of $Gl(n, \mathbb{R})$ then $\Pi_G^1(M, M)$ is locally diffeomorphic to $\mathbb{R}^n \times \mathbb{R}^n \times G$ if and only if $\Pi_G^1(M, M)$ locally diffeomorphic to $\mathbb{R}^n \times \mathbb{R}^n \times \tilde{G}$.

There is a special reduced subgroupoid of $\Pi^1(M, M)$ which will play an important role in the following. A trivial reduced subgroupoid of $\Pi^1(M, M)$ or *parallelism of* $\Pi^1(M, M)$ is a reduced subgroupoid of $\Pi^1(M, M)$, $\Pi_e^1(M, M) \rightrightarrows M$, such that for each $x, y \in M$ there exists a unique 1-jet $j_{x,y}^1 \phi \in \Pi_e^1(M, M)$.

So, having a trivial reduced subgroupoid of $\Pi^1(M, M)$, $\Pi_e^1(M, M)$, we can consider a map $\mathcal{P} : M \times M \to \Pi^1(M, M)$ such that $\mathcal{P}(x, y)$ is the unique 1-jet from $x$ to $y$ which is in $\Pi_e^1(M, M)$. It is easy to prove that $\mathcal{P}$ is, indeed, a global section of $(\alpha, \beta)$. Conversely, every global section of $(\alpha, \beta)$ (understanding "section" as section in the category of Lie groupoids, i.e., Lie groupoid morphism from the pair groupoid $M \times M$ to $\Pi^1(M, M)$ which is a section of the morphism $(\alpha, \beta)$) can be seen as a parallelism of $\Pi^1(M, M)$. Using this, we can also speak about *integrable sections of* $(\alpha, \beta)$.

Now, using the induced coordinates given in Eq. (3.3)

$$\Pi^1(U, V) : (x^i, y^j, y_i^j), \tag{5.5}$$

an integrable section can be written locally as follows,

$$\mathcal{P}(x^i, y^j) = (x^i, y^j, \delta_i^j),$$

or equivalently

$$\mathcal{P}(x, y) = j_{x,y}^1 (\psi^{-1} \circ \tau_{\psi(y) - \varphi(x)} \circ \varphi), \tag{5.6}$$

for some two local charts $(\varphi, U), (\psi, V)$ on $M$.

Notice that, using Proposition 5.1.2, $\mathcal{P}$ is an integrable section if and only if we can cover $M$ by local charts $(\varphi, U)$ such that

$$\mathcal{P}_{|U}(x, y) = j_{x,y}^1 (\varphi^{-1} \circ \tau_{\varphi(y) - \varphi(x)} \circ \varphi). \tag{5.7}$$

Next, analogously to the case of $G$-structures, we can characterize the integrable subgroupoids using (local) integrable sections (see Proposition C.2.3). However, in this case it is not so easy because, having a reduced subgroupoid, we do not know anything about the structure group $G$. So, firstly, we will have to solve this problem. Let $\Pi^1_G(M,M)$ be a reduced subgroupoid of $\Pi^1(M,M)$ and $\overline{z}_0 \in FM$ be a frame at $z_0 \in M$. Then, we define

$$G := \{\overline{z}_0^{-1} \cdot g \cdot \overline{z}_0 \ / \ g \in \Pi^{1\,z_0}_{G\,z_0}\} = \overline{z}_0^{-1} \cdot \Pi^{1\,z_0}_{G\,z_0} \cdot \overline{z}_0, \qquad (5.8)$$

where $\Pi^{1\,z_0}_{G\,z_0}$ is the isotropy group of $\Pi^1(M,M)$ at $z_0$. Therefore, $G$ is a Lie subgroup of $Gl(n,\mathbb{R})$. This Lie group will be called *associated Lie group* to $\Pi^1_G(M,M)$.

Note that, as a difference with $G$-structures, we do not have a unique Lie group $G$. In fact, let $\overline{y}_0$ be a frame at $y_0$ and $\tilde{G}$ be the associated Lie group, then, if we take $L_{z_0,y_0} \in \Pi^1_G(z_0,y_0)$ we have

$$G = [\overline{y}_0^{-1} \cdot L_{z_0,y_0} \cdot \overline{z}_0]^{-1} \cdot \tilde{G} \cdot [\overline{y}_0^{-1} \cdot L_{z_0,y_0} \cdot \overline{z}_0],$$

i.e., $G$ and $\tilde{G}$ are conjugated subgroups of $Gl(n,\mathbb{R})$. Notice that this fact is what we have expected because of Remark 5.1.3.

**Proposition 5.1.4.** *A reduced subgroupoid $\Pi^1_G(M,M)$ of $\Pi^1(M,M)$ is integrable if and only if for each two points $x,y \in M$ there exist coordinate systems $(x^i)$ and $(y^j)$ over $U,V \subseteq M$, respectively with $x \in U$ and $y \in V$ such that the local section,*

$$\mathcal{P}(x^i,y^j) = (x^i,y^j,\delta^j_i), \qquad (5.9)$$

*takes values into $\Pi^1_G(M,M)$.*

**Proof.** First, it is obvious that if $\Pi^1_G(M,M)$ is integrable then, we can restrict the maps $\Psi^{-1}_{U,V}$ to $\overline{U} \times \overline{V} \times \{e\}$ to get (local) integrable sections of $(\alpha,\beta)$ which takes values on $\Pi^1_G(M,M)$.

Conversely, in a similar way to Proposition 5.1.2 we can claim that for each $x \in M$ there exists an open set $U \subseteq M$ with $x \in U$ and $\mathcal{P} : U \times U \to \Pi^1_G(U,U)$ an integrable section of $(\alpha,\beta)$ given by

$$\mathcal{P}(x,y) = j^1_{x,y}\left(\psi_U^{-1} \circ \tau_{\psi_U(y)-\psi_U(x)} \circ \psi_U\right),$$

where $\psi_U : U \to \overline{U}$ is a local chart at $x$.

Then, we can construct the map

$$\Psi^{-1}_{U,U} : \overline{U} \times \overline{U} \times \{e\} \to \Pi^1_G(U,U),$$

defined in the obvious way.

Now, let $z_0 \in U$ be a point at $U$, $\overline{z}_0 = j^1_{0,z_0}\left(\psi_U^{-1} \circ \tau_{\psi_U(z_0)}\right) \in FU$ be a frame at $z_0$ and $G$ be the Lie subgroup satisfying Eq. (5.8). Then, we can define

$$\Psi_{U,U} : \Pi^1_G(U,U) \to \overline{U} \times \overline{U} \times G,$$

where for each $j^1_{z_0,z_0}\phi \in \Pi^1_G(z_0)$ and $x,y \in \overline{U}$ we define

$$\Psi_{U,U}^{-1}\left(x, y, Z_0^{-1} \cdot j^1_{z_0,z_0}\phi \cdot Z_0\right)$$
$$= j^1_{0,\psi_U^{-1}(y)}\left(\psi_U^{-1} \circ \tau_y\right) \cdot [\overline{z}_0^{-1} \cdot j^1_{z_0,z_0}\phi \cdot \overline{z}_0] \cdot j^1_{\psi_U^{-1}(x),0}\left(\tau_{-x} \circ \psi_U\right).$$

Hence the map $\Psi_{U,U} : \overline{U} \times \overline{U} \times G \to \Pi^1_G(U,U)$ is an isomorphism of Lie groupoids induced by $\psi_U$.

To end the proof, we only have to use Proposition 5.1.2.  □

Let $\mathcal{B}$ be a body. Taking into account the definition of homogeneity (see Definition 2.0.8) and the above result we can give the following proposition.

**Proposition 5.1.5.** *Let $\mathcal{B}$ be a uniform body. If $\mathcal{B}$ is homogeneous then $\Omega(\mathcal{B})$ is integrable. Conversely, $\Omega(\mathcal{B})$ is integrable implies that $\mathcal{B}$ is locally homogeneous.*

Now, we want to work with the notion of integrability in the associated Lie algebroid of the 1-jets groupoid. So, we will introduce this notion and relate it with the integrability of reduced subgroupoids of $\Pi^1(M,M)$.

Note that the induced map of the Lie groupoid isomorphism $L : \Pi^1(\mathbb{R}^n, \mathbb{R}^n) \to \mathbb{R}^n \times \mathbb{R}^n \times Gl(n, \mathbb{R})$ is given by a Lie algebroid isomorphism

$$AL : A\Pi^1(\mathbb{R}^n, \mathbb{R}^n) \to T\mathbb{R}^n \oplus (\mathbb{R}^n \times \mathfrak{gl}(n, \mathbb{R})),$$

where $T\mathbb{R}^n \oplus (\mathbb{R}^n \times \mathfrak{gl}(n, \mathbb{R}))$ is the trivial Lie algebroid on $\mathbb{R}^n$ with structure algebra $\mathfrak{gl}(n, \mathbb{R})$ (see Example 4.0.8).

Now, if $\mathfrak{g}$ is a Lie subalgebra of $\mathfrak{gl}(n, \mathbb{R})$, we can transport $T\mathbb{R}^n \oplus (\mathbb{R}^n \times \mathfrak{g})$ by this isomorphism to obtain a reduced Lie subalgebroid of $A\Pi^1(\mathbb{R}^n, \mathbb{R}^n)$. This kind of reduced subalgebroid will be called *standard flat* on $A\Pi^1(\mathbb{R}^n, \mathbb{R}^n)$.

Let $U \subseteq M$ be an open subset of $M$. We denote by $A\Pi^1(U, U)$ the open Lie subalgebroid of $A\Pi^1(M, M)$ defined by the associated Lie algebroid of $\Pi^1(U, U)$.

**Definition 5.1.6.** Let $A\Pi^1_G(M, M)$ be a Lie subalgebroid of $A\Pi^1(M, M)$. $A\Pi^1_G(M, M)$ is said to be *integrable by $G$* if it is locally isomorphic to the algebroid $T\mathbb{R}^n \oplus (\mathbb{R}^n \times \mathfrak{g})$, where $\mathfrak{g}$ is the Lie algebra of the Lie subgroup $G$ of $Gl(n, \mathbb{R})$.

Again, we need to explain what we understand by "locally isomorphic" in this case. $A\Pi^1_G(M, M)$ is locally diffeomorphic to $T\mathbb{R}^n \oplus (\mathbb{R}^n \times \mathfrak{g})$ if for all $x \in M$ there exists an open set $U \subseteq M$ with $x \in U$ and a local chart, $\psi_U : U \to \overline{U}$, which induces an isomorphism of Lie algebroids,

$$A\Psi_{U,U} : A\Pi^1_G(U, U) \to T\mathbb{R}^n \oplus (\mathbb{R}^n \times \mathfrak{g}), \tag{5.10}$$

such that $A\Psi_{U,U}$ is the induced map of the following isomorphism of Lie groupoids

$$\Psi_{U,U} : \Pi^1_G(U, U) \to \overline{U} \times \overline{U} \times G,$$

for some Lie subgroupoid $\Pi^1_G(U, U)$ of $\Pi^1(U, U)$, where for all $j^1_{x,y}\phi \in \Pi^1_G(U, U)$ the image $\Psi_{U,U}(j^1_{x,y}\phi)$ is given by

$$\left( \psi_U(x), \psi_U(y), j^1_{0,0} \left( \tau_{-\psi_U(y)} \circ \psi_U \circ \phi \circ \psi_U^{-1} \circ \tau_{\psi_U(x)} \right) \right), \tag{5.11}$$

So, for each open $U \subseteq M$, $A\Pi^1_G(U, U)$ is integrable by a Lie subgroupoid $\Pi^1_G(U, U)$ of $\Pi^1(U, U)$. Using the uniqueness of integrating immersed (source-connected) subgroupoids (see Proposition 4.0.40), $A\Pi^1_G(M, M)$ is integrable by a Lie subgroupoid of $\Pi^1(M, M)$ which will be denoted by $\Pi^1_G(M, M)$.

Notice that, in this case, it does not make sense to use two different open sets $U$ and $V$ like in Definition 5.1.1 and Eq. (5.1). This is because, if $U$ and $V$ were different, the set $\Pi^1_G(U, U)$ would not be a groupoid and then, the morphism $\Psi_{U,V}$ does not have to induce a morphism of Lie algebroids.

By definition, it is immediate to prove that $A\Pi_G(M, M)$ is integrable by $G$ if and only if, $\Pi_G(M, M)$ is integrable (by using Proposition 5.1.2).

Analogously to the case of 1-jets groupoid, a *parallelism* of $A\Pi^1(M, M)$ is an associated Lie algebroid of a parallelism of

$\Pi^1(M, M)$. Hence, using the Lie's second fundamental theorem (Theorem 4.0.39), a parallelism is a section of $\sharp$ (understanding "section" as section in the category of Lie algebroids, i.e., Lie algebroid morphism from the tangent algebroid $TM$ to $A\Pi^1(M, M)$ which is a section of the morphism $\sharp$) and reciprocally. In this way, we will also speak about *integrable sections of* $\sharp$.

Let $(x^i)$ be a local coordinate system defined on some open subset $U \subseteq M$. Then, we will use the local coordinate system defined in Eq. (4.18),

$$A\Pi^1(U, U) : \left( \left( x^i, x^i, \delta_j^i \right), 0, v^i, v_j^i \right) \cong \left( x^i, v^i, v_j^i \right),$$

which are, indeed, induced coordinates by the functor $A$ from local coordinates on $\Pi^1(U, U)$.

Notice that each integrable section of $(\alpha, \beta)$ in $\Pi^1(M, M)$, $\mathcal{P}$, is a Lie groupoid morphism. Hence, $\mathcal{P}$ induces a Lie algebroid morphism $A\mathcal{P} : TM \to A\Pi^1(M, M)$ (see Theorem 4.0.26) which is a section of $\sharp$ and is given by

$$A\mathcal{P}(v_x) = T_x \mathcal{P}_x(v_x), \quad \forall v_x \in T_x M,$$

where $\mathcal{P}_x : M \to \Pi_x^1(M, M)$ satisfies that

$$\mathcal{P}_x(y) = \mathcal{P}(x, y), \quad \forall x, y \in M.$$

So, taking into account that, locally,

$$\mathcal{P}\left( x^i, y^j \right) = (x^i, y^j, \delta_i^j),$$

we have that each integrable section can be written locally as follows:

$$A\mathcal{P}\left( x^i, \frac{\partial}{\partial x^i} \right) = \left( x^i, \frac{\partial}{\partial x^i}, 0 \right).$$

Now, using Proposition 5.1.4, we have the following analogous proposition.

**Proposition 5.1.7.** *A reduced subalgebroid $A\Pi_G^1(M, M)$ of $A\Pi^1(M, M)$ is integrable by $G$ if and only if there exist local integrable sections of $\sharp$ covering $M$ which takes values in $A\Pi_G^1(M, M)$.*

Equivalently, for each point $x \in M$ there exists a local coordinate system $\left(x^i\right)$ over an open set $U \subseteq M$ with $x \in U$ such that the local sections

$$\Delta\left(x^i, \frac{\partial}{\partial x^i}\right) = \left(x^i, \frac{\partial}{\partial x^i}, 0\right),$$

take values in $A\Pi^1_G\left(M, M\right)$.

Finally, we will use the algebroid of derivations on $M$. Thus, as we have shown in Chapter 4, the map $\mathcal{D} : \Gamma\left(A\Pi^1\left(M, M\right)\right) \to Der\left(TM\right)$ given by

$$\mathcal{D}\left(\Lambda\right) = D^\Lambda = D^\Lambda = \frac{\partial}{\partial t_{|0}}\left(\varphi_t^\Lambda\right)^*,$$

defines a Lie algebroid isomorphism $\mathcal{D} : A\Pi^1\left(M, M\right) \to \mathfrak{D}\left(TM\right)$ over the identity map on $M$.

Let $\Delta$ be a linear section of $\sharp$ in $A\Pi^1\left(M, M\right)$. Then, $\mathcal{D}$ induces a covariant derivative on $M$, $\nabla^\Delta$. Thus, for each $\left(x^i\right)$ local coordinate system on $M$,

$$\Delta\left(x^i, \frac{\partial}{\partial x^j}\right) = \left(x^i, \frac{\partial}{\partial x^j}, \Delta^k_{i,j}\right).$$

Hence, remember that the functions $\Delta^k_{i,j}$ are the Christoffel symbols of $\nabla^\Delta$, i.e.,

$$\nabla^\Delta_{\frac{\partial}{\partial x^j}} \frac{\partial}{\partial x^i} = \Delta^k_{i,j} \frac{\partial}{\partial x^k}.$$

With this fact in mind, we can give another characterization of the integrability over the 1-jets algebroid.

**Proposition 5.1.8.** *Let $\Delta$ be a linear section of $\sharp$ in the 1-jets Lie algebroid, $A\Pi^1\left(M, M\right)$. Then, it is integrable if and only if for each point $x \in M$ there exists a local coordinate system $\left(x^i\right)$ on an open set $U \subseteq M$ with $x \in U$ such that $\nabla^\Delta$ is a covariant derivative with Christoffel symbols equal to zero, i.e.,*

$$\nabla^\Delta_{\frac{\partial}{\partial x^j}} \frac{\partial}{\partial x^i} = 0, \quad \forall i, j.$$

In other words, integrable linear section of $\natural$ coincide with the torsion-free and flat connections (see Appendix B, Lemma B.0.13).

Let $W : \Pi^1(\mathcal{B}, \mathcal{B}) \to V$ be the mechanical response which defines $\Omega(\mathcal{B})$. Consider a section $\Lambda \in \Gamma(A\Omega(\mathcal{B}))$. So, the flow of the left-invariant vector field $\Theta^\Lambda$, $\{\varphi_t^\Lambda\}$, can be restricted to $\Omega(\mathcal{B})$. Hence, we have

$$W\left(\varphi_t^\Lambda(g)\right) = W\left(\varphi_t^\Lambda\left(g \cdot \bar{\epsilon}\left(\bar{\alpha}\left(g\right)\right)\right)\right)$$
$$= W\left(g \cdot \varphi_t^\Lambda\left(\bar{\epsilon}\left(\bar{\alpha}\left(g\right)\right)\right)\right)$$
$$= W\left(g\right)$$

for all $g \in \Pi^1(\mathcal{B}, \mathcal{B})$. Thus, for each $g \in \Pi^1(\mathcal{B}, \mathcal{B})$, we deduce

$$TW\left(\Theta^\Lambda(g)\right) = \frac{\partial}{\partial t_{|0}}\left(W\left(\varphi_t^\Lambda(g)\right)\right) = \frac{\partial}{\partial t_{|0}}\left(W(g)\right) = 0.$$

Therefore,

$$TW\left(\Theta^\Lambda\right) = 0. \tag{5.12}$$

Conversely, it is easy to prove that Eq. (5.12) implies that $\Theta \in \Gamma\left(A\overline{\Omega}(\mathcal{B})\right)$.

In this way, the material algebroid can be defined without using the material groupoid by imposing Eq. (5.12). Thus, we can characterize the homogeneity and uniformity using the material Lie algebroid.

Now, from the above results we can give the following ones.

**Proposition 5.1.9.** *Let $\mathcal{B}$ be a uniform body. If $\mathcal{B}$ is homogeneous, then, $A\Omega(\mathcal{B})$ is integrable by a Lie subgroup $G$ of $Gl(n, \mathbb{R})$. Conversely, if $A\Omega(\mathcal{B})$ is integrable by $G$ then $\mathcal{B}$ is locally homogeneous.*

Using Proposition 5.1.7, this result can be expressed locally as follows.

**Proposition 5.1.10.** *Let $\mathcal{B}$ be a uniform body. $\mathcal{B}$ is locally homogeneous if and only if for each point $x \in \mathcal{B}$ there exists a local coordinate system $\left(x^i\right)$ over $U \subseteq \mathcal{B}$ with $x \in U$ such that the local section of $\natural$,*

$$\Delta\left(x^i, \frac{\partial}{\partial x^i}\right) = \left(x^i, \frac{\partial}{\partial x^i}, 0\right),$$

*takes values in $A\Omega(\mathcal{B})$.*

Finally, denoting by $\mathcal{D}(\mathcal{B})$ to the Lie subalgebroid of the derivation algebroid on $\mathcal{B}$, $\mathcal{D}(A\Omega(\mathcal{B})) \leq \mathfrak{D}(T\mathcal{B})$, we can give the following result.

**Theorem 5.1.11.** *Let $\mathcal{B}$ be a uniform body. If $\mathcal{B}$ is homogeneous respect to the global deformation $\kappa$, there exists a global covariant derivative on $\mathcal{B}$ which takes values in $\mathcal{D}(\mathcal{B})$ and is trivial respect to $\kappa$.*

*Conversely, $\mathcal{B}$ is locally homogeneous if and only if for each point $x \in \mathcal{B}$ there exists a local coordinate systems $(x^i)$ over $U \subseteq \mathcal{B}$ with $x \in U$ such that the local covariant derivative on $\mathcal{B}$ characterized by,*

$$\nabla_{\frac{\partial}{\partial x^j}} \frac{\partial}{\partial x^i} = 0, \ \forall i, j,$$

*which takes values in $\mathcal{D}(\mathcal{B})$.*

Roughly speaking, $\mathcal{B}$ is locally homogeneous if and only if there exist local trivial covariant derivatives on $\mathcal{B}$ which take values in $\mathcal{D}(\mathcal{B})$.

**Remark 5.1.12.** There is still another interesting way of interpreting the 1-jets Lie groupoid on a body $\mathcal{B}$ (and, hence, of interpreting the integrability of a reduced subgroupoid $\Pi_G^1(\mathcal{B},\mathcal{B})$ of $\Pi^1(\mathcal{B},\mathcal{B})$). As we know, there exists another structure of Lie groupoid related with $F\mathcal{B}$, the Gauge groupoid of the principal bundle $F\mathcal{B}$ (see Example 3.0.21).

We only have to take into account that

$$\text{Gauge}(F\mathcal{B}) \cong \mathcal{B} \times F\mathcal{B}.$$

Furthermore, translating points we can construct an isomorphism of Lie groupoid from $\mathcal{B} \times F\mathcal{B}$ to $\Pi^1(\mathcal{B},\mathcal{B})$ (notice that this isomorphism depends on the reference configuration $\phi_0$). Thus, the 1-jets Lie groupoid can be seen as the gauge groupoid of the principal bundle $F\mathcal{B}$ and, therefore, the 1-jets Lie algebroid can be seen as the Atiyah algebroid associated with $F\mathcal{B}$.

**Remark 5.1.13.** Notice that, using Eq. (5.12) we can characterize the Lie subalgebroid $\mathcal{D}(\mathcal{B})$ of the derivation algebroid on $\mathcal{B}$ by the derivations on $\mathcal{B}$, $D$, such that the associated section of $A\Pi^1(\mathcal{B},\mathcal{B})$ satisfies Eq. (5.12).

Let $(x^i)$ be a local coordinate system on $\mathcal{B}$ and $D$ be a derivation on $\mathcal{B}$ with base vector field $\Theta$. We denote

- $\Theta\left(x^i\right) = \left(x^i, \Theta^j\right)$.
- $D\left(\frac{\partial}{\partial x^i}\right) = \Theta_i^j \frac{\partial}{\partial x^j}$.

Then, $D$ is in $\mathcal{D}(\mathcal{B})$ if and only if over any $(x^i)$ local coordinate system on $\mathcal{B}$ it is satisfied that

$$dW_{|\left(x^i, x^i, g_i^j\right)}(0, \Theta^j, g_l^j \cdot \Theta_i^l) = 0,$$

for all material symmetry $g \in G(x)$ which is locally written as follows

$$g \cong (x^i, x^i, g_i^j).$$

## 5.2   Homogeneity with $G$-structures

Finally, we will prove that our definition of homogeneity (see Definition 2.0.8 or Proposition 5.1.5) is, indeed, equivalent to that used in Elżanowski *et al.* (1990) where the authors use $G$-structures to characterize this property (see Definition 2.0.11).

Let $\overline{Z}_0$ be a fixed frame at $Z_0 \in \mathcal{B}$. Then, we construct a $G_0$-structure $\omega_{G_0}(\mathcal{B})$ on $\mathcal{B}$ containing $\overline{Z}_0$ given by

$$\omega_{G_0}(\mathcal{B}) = \Omega_{Z_0}(\mathcal{B}) \cdot \overline{Z}_0.$$

So, Proposition 2.0.12 shows us that the (local) homogeneity with respect to $\overline{Z}_0$ is equivalent to the integrability of $\omega_{G_0}(\mathcal{B})$.

To compare both definitions, we will start constructing the following map:

$$\mathcal{G} : \Gamma(FM) \to \Gamma_{(\alpha,\beta)}\left(\Pi^1(M, M)\right)$$
$$P \mapsto \mathcal{G}P.$$

such that

$$\mathcal{G}P(x, y) = P(y) \cdot [P(x)^{-1}], \tag{5.13}$$

where $\cdot$ is the composition of 1-jets. Obviously, $\mathcal{G}$ is well-defined.

Before starting to work with integrable sections, we are interested in dilucidating when an element of $\Gamma_{(\alpha,\beta)}\left(\Pi^1(M, M)\right)$ can be inverted by $\mathcal{G}$. First, we consider $P \in \Gamma(FM)$; then for all $x, y, z \in M$, we have

$$\mathcal{G}P(y, z) \cdot \mathcal{G}P(x, y) = \mathcal{G}P(x, z), \tag{5.14}$$

i.e., $\mathcal{G}P$ is a morphism of Lie groupoids over the identity map on $M$ from the pair groupoid $M \times M$ to $\Pi^1(M, M)$. Therefore, not every element of $\Gamma_{(\alpha, \beta)}(\Pi^1(M, M))$ can be inverted by $\mathcal{G}$ but we can prove the following result.

**Proposition 5.2.1.** *Let $\mathcal{P}$ be a section of $(\alpha, \beta)$ in $\Pi^1(M, M)$. Then there exists $P$ a section of $FM$ such that*

$$\mathcal{G}P = \mathcal{P},$$

*if and only if $\mathcal{P}$ is a morphism of Lie groupoids over the identity map from the pair groupoid $M \times M$ to $\Pi^1(M, M)$.*

**Proof.** We have proved that the existence of such $P$ implies that $\mathcal{P}$ is morphism of Lie groupoids. Conversely, if Eq. (5.14) is satisfied we can define $P \in \Gamma(FM)$ as follows:

$$P(x) = \mathcal{P}(z, x) \cdot j_{0,z}^1 \psi,$$

where $j_{0,z}^1 \psi \in FM$ is fixed. Then, using Eq. (5.14), we have

$$\mathcal{G}P = \mathcal{P}. \qquad \square$$

However, there is not a unique $P$ such that $\mathcal{G}P = \mathcal{P}$. We will study this problem in Remark 5.2.2. Notice that the relevant sections of $(\alpha, \beta)$ are going to be the parallelisms which are, indeed, the morphisms of Lie groupoids over the identity map from the pair groupoid $M \times M$ to $\Pi^1(M, M)$.

Next, suppose that $P$ is an integrable section of $FM$. Then, for each point $x \in M$ there exists a local coordinate system $(x^i)$ on $M$ such that

$$P(x^i) = (x^i, \delta_j^i),$$

or equivalently,

$$P(x) = j_{0,x}^1 \left( \varphi^{-1} \circ \tau_{\varphi(x)} \right), \tag{5.15}$$

where $\varphi$ is the local chart over $x$ and $\tau_{\varphi(x)}$ denote the translation on $\mathbb{R}^n$ by the vector $\varphi(x)$.

Then, for all $x, y \in M$ there exist two charts $\psi$ and $\varphi$ over $x$ and $y$ respectively such that

$$\mathcal{G}P\left(x, y\right) = j_{x,y}^1\left(\psi^{-1} \circ \tau_{\psi(y) - \varphi(x)} \circ \varphi\right), \qquad (5.16)$$

i.e., $\mathcal{G}P$ is an integrable section of $(\alpha, \beta)$ on $\Pi^1\left(M, M\right)$. Hence $\mathcal{G}$ applies integrable sections on $FM$ into integrable sections in $\Pi^1\left(M, M\right)$. Furthermore, using Eq. (5.6), for each integrable section of $(\alpha, \beta)$ $\mathcal{P}$ in $\Pi^1\left(M, M\right)$ we can construct $P$, integrable section on $FM$ such that

$$\mathcal{G}P = \mathcal{P}.$$

However, this fact does not implies that $P$ integrable is equivalent to $\mathcal{G}P$ integrable. So, we will study this problem in the following remark.

**Remark 5.2.2.** Let $P, Q : M \to FM$ be two sections of $FM$ such that $\mathcal{G}P = \mathcal{G}Q$, i.e., for all $x, y \in M$, we have

$$P\left(y\right) \cdot \left[P\left(x\right)^{-1}\right] = Q\left(y\right) \cdot \left[Q\left(x\right)\right]^{-1}.$$

Then,

$$\left[Q\left(y\right)\right]^{-1} \cdot P\left(y\right) = \left[Q\left(x\right)\right]^{-1} \cdot P\left(x\right).$$

So, denoting by $Z_0 = \left[Q\left(x\right)\right]^{-1} \cdot P\left(x\right)$, we deduce that

$$P\left(x\right) = Q\left(x\right) \cdot Z_0.$$

Conversely, for each 1-jet $Z_0 = j_{0,0}^1\phi \in F\mathbb{R}_0^n$, where $F\mathbb{R}_0^n$ is the fibre of $F\mathbb{R}^n$ over 0, and each section of $FM$, $P : M \to FM$, the section of $FM$ given by

$$Q\left(x\right) = P\left(x\right) \cdot Z_0, \qquad (5.17)$$

satisfies that

$$\mathcal{G}P = \mathcal{G}Q.$$

Thus, we have shown that for each section on $FM$, $P : M \to FM$

$$\mathcal{G}^{-1}\left(\mathcal{G}P\right) = \{P \cdot Z_0 \ / \ Z_0 \in F\mathbb{R}_0^n\},$$

i.e., the map $\mathcal{G}$ can be considered as an injective map over the quotient space by Eq. (5.17).

Using this, it is obvious that, if $\mathcal{G}P$ is integrable, then, $P$ is integrable too, i.e., the map $\mathcal{G}$ restricted to the integrable sections can be considered as a one-to-one map over the quotient space by Eq. (5.17).

Finally, we can generalize the map $\mathcal{G}$ into a map which applies $G$-structures on $M$ into reduced subgroupoids of $\Pi^1(M,M)$. Let $\omega_G(M)$ be a $G$-structure on $M$, then we consider the following set:

$$\mathcal{G}(\omega_G(M)) = \{L_y \cdot [L_x^{-1}] \ / \ L_x, L_y \in \omega_G(M)\}.$$

It is straightforward to prove that $\mathcal{G}(\omega_G(M))$ is a reduced subgroupoid of $\Pi^1(M,M)$. In fact, taking a local section of $\omega_G(M)$,

$$P_U : U \to \omega_G(U),$$

the map given by

$$F_U : \Pi^1(U,U) \to FU$$
$$L_{x,y} \mapsto L_{x,y} \cdot [P_U(x)]$$

is a diffeomorphism which satisfies that $F_U\left(\Pi_G^1(U,U)\right) = \omega_G(U)$.

Analogously to parallelisms, we can prove that every reduced subgroupoid can be inverted by $\mathcal{G}$ into a $G$-structure on $M$, where $G$ is defined by Eq. (5.8) with $Z_0 \in FM$ fixed.

We consider $z_0 = \pi_M(Z_0)$. Then, we can generate a $G$-structure over $M$ in the following way:

$$\omega_G(M) := \{L_{z_0,x} \cdot Z_0 \cdot g \ / \ g \in G, \ L_{z_0,x} \in \Pi_G^1(M,M)_{z_0}\}.$$

Notice that the fibre of $\omega_G(M)$ at $x \in M$ is given by the set

$$\{L_{z_0,x} \cdot Z_0 \cdot g \ / \ g \in G\},$$

for any fixed $L_{z_0,x} \in \Pi_G^1(M,M)_{z_0}$. In fact, for two $L_{z_0,x}, G_{z_0,x} \in \Pi_G^1(M,M)_{z_0}$,

$$[L_{z_0,x} \cdot Z_0]^{-1} \cdot G_{z_0,x} \cdot Z_0 \in G.$$

Notice that the map $L_{z_0,x} \to L_{z_0,x} \cdot Z_0$ defines an isomorphism of principal bundles from $\Pi_G^1(M,M)_{z_0}$ to $\omega_G(M)$.

Finally, let $\omega_G(M)$ be an integrable $G$-structure on $M$, then using Proposition C.2.3 and Eq. (5.15), for each point $x \in M$ there exists a local chart $(\varphi, U)$ with $x \in U$ such that $\omega_G(U)$ is given by the 1-jets $j_{0,x}^1\left(\varphi^{-1} \circ \tau_{\varphi(x)}\right) \cdot A$ for all $x \in U$ and $A \in G$.

Therefore, taking two local charts $(\varphi, U)$ and $(\psi, V)$ and denoting $\mathcal{G}(\omega_G(M))$ by $\Pi^1_G(M, M)$ we have that the elements of $\Pi^1_G(U, V)$ are given by

$$j^1_{0,y}\left(\psi^{-1} \circ \tau_{\psi(y)}\right) \cdot A \cdot j^1_{x,0}\left(\tau_{-\varphi(x)} \circ \varphi\right) \tag{5.18}$$

for all $x \in U$, $y \in V$, $A \in G$. So, the local section of $(\alpha, \beta)$ given by $j^1_{x,y}\left(\psi^{-1} \circ \tau_{\psi(y)-\varphi(x)} \circ \varphi\right)$ is in $\Pi_G(M, M)$, i.e., $\Pi_G(M, M)$ is integrable.

Next, to prove the converse we only have to construct $\omega_G(M)$ using Eq. (5.18) and repeat the above construction of a $G$-structure which inverts $\Pi^1_G(M, M)$.

**Remark 5.2.3.** Let $\omega_G(M)$ be a $G$-structure on $M$ and $\overline{\omega}_{\overline{G}}(M)$ be a $\overline{G}$-structure on $M$ such that $\mathcal{G}(\omega_G(M)) = \mathcal{G}(\overline{\omega}_{\overline{G}}(M))$, i.e., for all $j^1_{0,x}\phi, j^1_{0,y}\theta \in \omega_G(M)$, there exist $j^1_{0,x}\overline{\phi}, j^1_{0,y}\overline{\theta} \in \overline{\omega}_{\overline{G}}(M)$ such that

$$j^1_{x,y}\left(\theta \circ \phi^{-1}\right) = j^1_{x,y}(\overline{\theta} \circ \overline{\phi}^{-1}).$$

Then,

$$j^1_{0,0}\left(\theta^{-1} \circ \overline{\theta}\right) = j^1_{0,0}\left(\phi^{-1} \circ \overline{\phi}\right).$$

So, denoting by $Z_0 = j^1_{0,0}\left(\phi^{-1} \circ \overline{\phi}\right)$, we have

$$\omega_G(M) \cdot Z_0 = \overline{\omega}_{\overline{G}}(M), \tag{5.19}$$

In fact, for $j^1_{0,x}\psi \in \omega_G(M)$ we have that

$$\left(j^1_{0,x}\psi\right) \cdot Z_0 = j^1_{x,x}\left(\psi \circ \phi^{-1}\right) \cdot j^1_{0,x}\overline{\phi} \in \overline{\omega}_{\overline{G}}(M),$$

taking into account that

$$j^1_{x,x}\left(\psi \circ \phi^{-1}\right) \in \mathcal{G}\left(\overline{\omega}_{\overline{G}}(M)\right).$$

Hence,

$$\omega_G(M)_x \cdot Z_0 \subseteq \overline{\omega}_{\overline{G}}(M)_x.$$

The converse is proved in the same way and, so

$$\omega_G(M)_x \cdot Z_0 = \overline{\omega}_{\overline{G}}(M)_x. \tag{5.20}$$

Finally, in general, if Eq. (5.20) is satisfied for one point it is easy to prove that

$$\omega_G(M) \cdot Z_0 = \overline{\omega}_{\overline{G}}(M).$$

Note that this implies that the isotropy groups are conjugate, namely

$$\overline{G} = Z_0 \cdot G \cdot Z_0^{-1}.$$

This kind of $G$-structure is called *conjugated $G$-structure*.

Conversely, for all $\omega_G(M), \overline{\omega}_{\overline{G}}(M)$, conjugated $G$-structures, we have

$$\mathcal{G}(\omega_G(M)) = \mathcal{G}(\overline{\omega}_{\overline{G}}(M)).$$

Using this, if $\mathcal{G}(\omega_G(M))$ is integrable, then $\omega_G(M)$ is integrable too.

Let $\mathcal{B}$ be a smoothly uniform body. Using the above results, the $G_0$-structure $\omega_{G_0}(\mathcal{B})$ is integrable if and only if $\mathcal{G}(\omega_{G_0}(\mathcal{B}))$ is integrable.

Furthermore, it is clear by construction that

$$\mathcal{G}(\omega_{G_0}(\mathcal{B})) = \Omega(\mathcal{B}).$$

Therefore, using Propositions 5.1.5 and 2.0.12, we have effectively proved that both definitions are equivalent.

## Example

We will introduce the following example, based on the model of a so-called *simple liquid crystal*. These simple materials were introduced by Coleman (1965) and Wang (1965).

Let $\mathcal{B}$ be a simple body (we will assume that $\mathcal{B}$ is an open subset of $\mathbb{R}^3$ by taking the image by the reference configuration $\phi_0$) with a mechanical response $W : \Pi^1(\mathcal{B}, \mathcal{B}) \to V$ such that for all $h = j^1_{X,Y}\phi \in \Pi^1(\mathcal{B}, \mathcal{B})$ we have

$$W(h) = \widehat{W}(r(h), J(h)),$$

where, denoting by $F$ the associated matrix to $j^1_{X,Y}\phi$ (with respect to the canonical basis of $\mathcal{B}$),

- $r(j^1_{X,Y}\phi) = g(Y)(T_X\phi(e(X)), T_X\phi(e(X)))$;
- $J(j^1_{X,Y}\phi) = \det(F)$

with $e \in \mathfrak{X}(\mathcal{B})$ a vector field which is not zero at any point and $g$ a Riemannian metric on $\mathbb{R}^3$. Notice that the tangent bundle $T\mathcal{B}$ is canonically isomorphic to $\mathcal{B} \times \mathbb{R}^3$. So, for each $Y \in \mathcal{B}$, $g(Y)$ can be seen as an inner product on $\mathbb{R}^3$. Then, the expression of $r$ turns into the following,

$$r\left(j^1_{X,Y}\phi\right) = g(Y)\left(F \cdot \left(e^I(X)\right), F \cdot \left(e^I(X)\right)\right),$$

for all $X \in \mathcal{B}$, $e(X) = \left(X, e^I(X)\right)$, where we are using the canonical isomorphism $T\mathcal{B} \to \mathcal{B} \times \mathbb{R}^3$. We will use both expressions with the same notation.

Now, we want to study the condition which characterizes the material algebroid: A left-invariant vector field $\Theta \in \mathfrak{X}_L\left(\Pi^1(\mathcal{B}, \mathcal{B})\right)$ restricts to a section of $A\Omega(\mathcal{B})$ if and only if

$$\Theta(W) = 0.$$

So, we should study $TW$ over left-invariant vector fields. Let $\Theta \in \mathfrak{X}_L\left(\Pi^1(\mathcal{B}, \mathcal{B})\right)$ be a left-invariant vector field and consider the canonical local system of coordinates $\left(X^I\right)$ in $\mathbb{R}^3$ restricted to $\mathcal{B}$. Notice that, in fact, $\left(X^I\right)$ is the system of coordinates generated by the reference configuration $\phi_0$. We will denote by $\left(X^I, Y^J, F^J_I\right)$ the induced local coordinates of $\left(X^I\right)$ in $\Pi^1(\mathcal{B}, \mathcal{B})$. The local expression of $\Theta$ will be denoted as follows,

$$\Theta\left(X^I, Y^J, F^J_I\right) = \left(\left(X^I, Y^J, F^J_I\right), \delta X^I, 0, F^J_L \delta P^L_I\right).$$

Now, we will begin given the derivatives of $r$ and $J$. For each $A \in gl(3, \mathbb{R})$ and $v \in \mathbb{R}^3$ we have that,

(i) $\dfrac{\partial r}{\partial X^i_{|j^1_{X,Y}}}(v) = 2g(Y)\left(T_X\phi\left(e(X)\right), T_X\phi\left(\dfrac{\partial e}{\partial X_{|X}}(v)\right)\right),$

(ii) $\dfrac{\partial r}{\partial F^i_{|j^1_{X,Y}}}(A) = 2g(Y)\left(F \cdot \left(e^I(X)\right), A \cdot \left(e^I(X)\right)\right),$

(iii) $\dfrac{\partial J}{\partial F^i_{|j^1_{X,Y}}}(A) = \det(F)\, Tr\left(F^{-1} \cdot A\right).$

Here $F$ is the Jacobian matrix of $\phi$ at $X$ and $\dfrac{\partial e}{\partial X_{|X}}(v)$ is the tangent vector at $X$ such that

$$\frac{\partial e}{\partial X_{|X}}(v) = \left(X, \frac{\partial e^I}{\partial X^L_{|X}} v^L\right).$$

Hence, $\Theta$ restricts to a section of the material algebroid $A\Omega(\mathcal{B})$ if and only if

$$0 = 2\frac{\partial \widehat{W}}{\partial r_{|j_{X,Y}^1}} g(Y) \left( T_X \phi(e(X)), T_X \phi\left( \frac{\partial e}{\partial X_{|X}} (\delta X^i(X)) \right) \right)$$

$$+ 2\frac{\partial \widehat{W}}{\partial r_{|j_{X,Y}^1}} g(Y) \left( F \cdot (e^I(X)), F_i^j \delta P_i^l(X) \cdot (e^i(X)) \right)$$

$$+ \det(F) \frac{\partial \widehat{W}}{\partial J_{|j_{X,Y}^1}} \mathrm{Tr}(\delta P_i^j(X)),$$

for all $j_{X,Y}^1 \phi \in \Pi^1(\mathcal{B}, \mathcal{B})$. So, a sufficient but not necessary condition would be

**(1)** $\mathrm{Tr}\left( \delta P_I^J(X) \right) = 0.$

**(2)** Denoting by $\mathcal{L}_X = \frac{\partial e}{\partial X_{|X}} (\delta X^I(X)) + \delta P_I^J(X) \cdot (e^I(X))$, then

$$g(Y) \left( F \cdot (e^I(X)), F_M^R \cdot \mathcal{L}_X \right) = 0.$$

By using that $g$ is non-degenerate and $e(X)$ is non-zero, we turn these conditions into the following

**(1)'** $\delta P_I^I = 0,$

**(2)'** $\frac{\partial e^J}{\partial X^L} \delta X^L + \delta P_L^J e^L = 0, \ \forall J,$

where $e^J$ are the coordinates of $e$ respect to $(X^J)$.

Let us now study the uniformity of the material. By using proposition 4.0.47 $\mathcal{B}$ is uniform if and only if the material algebroid of $\mathcal{B}$ is transitive.

Let $V_X = (X, V^I)$ be a vector at $X \in \mathcal{B}$. Then, we should find a (local) left-invariant vector field $\Theta$ such that

- $\Theta(W) = 0,$
- $T_{\epsilon(X)} \alpha(\Theta(\epsilon(X))) = V_X,$

where $\epsilon$ and $\alpha$ are the identities map and the source map of the material groupoid, respectively.

Let us fix the local expression of $\Theta$ as follows,

$$\Theta\left(X^I, Y^J, F_I^J\right) = \left(\left(X^I, Y^J, F_I^J\right), \delta X^I, 0, F_L^J \delta P_I^L\right).$$

Then,

$$T_{\epsilon(X)}\alpha\left(\Theta\left(\epsilon\left(X\right)\right)\right) = \left(X^I\left(X\right), \delta X^I\left(X\right)\right).$$

So, it should satisfy that,

$$\delta X^I\left(X\right) = V^I, \quad \forall I.$$

By taking into account identities **(1)′** and **(2)′**, it is enough to find a family of (local) maps $A_i^j$ from the body to the space of matrices satisfying that

**(1)″** $A_I^I = 0$,

**(2)″** $\frac{\partial e^J}{\partial X^L} V^L = -A_L^J e^L, \ \forall J.$

It is just an easy exercise to prove that there are infinite solutions $A_I^J$ of the equations **(1)″** and **(2)″** and, hence, $\mathcal{B}$ is (smoothly) uniform.

From now on, we will assume that $\widehat{W}$ is an immersion. In that way, **(1)′** and **(2)′** are also necessary conditions.

Next, we will study the condition of (local) homogeneity. As we know (Proposition 5.1.10) $\mathcal{B}$ is (locally) homogeneous if and only if there exists a local system of coordinates $\left(x^i\right)$ such that the local section of $\sharp$,

$$\Delta\left(x^i, \frac{\partial}{\partial x^i}\right) = \left(x^i, \frac{\partial}{\partial x^i}, 0\right),$$

takes values in the material algebroid $A\Omega\left(\mathcal{B}\right)$. Equivalently,

$$\frac{\partial W}{\partial x^i} = 0, \quad \forall i. \tag{5.21}$$

So, let us study this equality. Notice that,

$$\frac{\partial W}{\partial x^i} = \frac{\partial \widehat{W}}{\partial r}\frac{\partial r}{\partial x^i} + \frac{\partial \widehat{W}}{\partial J}\frac{\partial J}{\partial x^i}.$$

Thus, by using that $\widehat{W}$ is an immersion, $(x^i)$ are homogeneous coordinates if and only if

**(1)'''** $\frac{\partial r}{\partial x^i} = 0$, $\forall i$,

**(2)'''** $\frac{\partial J}{\partial x^i} = 0$, $\forall i$.

Observe that the form of $\widehat{W}$ is not important to evaluate the (local) homogeneity of $\mathcal{B}$ as long as $\widehat{W}$ is an immersion.

Let $(x^i)$ be a system of homogeneous coordinates on $\mathcal{B}$. Then, for each $j^1_{X,Y}\phi \in \Pi^1(\mathcal{B},\mathcal{B})$

$$r\left(j^1_{X,Y}\phi\right) = g\left(Y\right)\left(T_X\phi\left(e\left(X\right)\right), T_X\phi\left(e\left(X\right)\right)\right)$$

$$= g\left(Y\right)\left(T_X\phi\left(e^i\left(X\right)\frac{\partial}{\partial x^i_{|X}}\right), \; T_X\phi\left(e^j\left(X\right)\frac{\partial}{\partial x^j_{|X}}\right)\right)$$

$$= e^i\left(X\right)e^j\left(X\right)\frac{\partial\phi^k}{\partial x^i_{|X}}\frac{\partial\phi^l}{\partial x^j_{|X}}g_{kl}(Y),$$

where, in this case, $e^j$ are the coordinates of $e$ respect to $(x^i)$. So, considering the induced coordinates $(x^i, y^j, y^j_i)$ of $(x^i)$ on $\Pi^1(\mathcal{B},\mathcal{B})$ we have that

$$r \circ (x^i, y^j, y^j_i)^{-1}(\tilde{X}, \tilde{Y}, \tilde{F}) = e^i\left(X\right)e^j\left(X\right)\tilde{F}^k_i\tilde{F}^l_j g_{kl}\left(Y\right),$$

for all $(\tilde{X}, \tilde{Y}, \tilde{F})$. In this way,

$$\frac{\partial r}{\partial x^k_{|j^1_{X,Y}}} = 2\frac{\partial e^i}{\partial x^k_{|X}}e^j\left(X\right)\tilde{F}^k_i\tilde{F}^l_j g_{kl}\left(Y\right).$$

Hence, by using the non-degeneracy of $g$ we have that $\frac{\partial r}{\partial x^k} = 0$ if and only if

$$\frac{\partial e^i}{\partial x^k} = 0, \quad \forall i. \tag{5.22}$$

Therefore, **(1)''** is satisfied if and only if the vector field $e$ is constant with respect to $(x^i)$, i.e.,

$$e = \lambda^i\frac{\partial}{\partial x^i}, \quad \lambda^i \equiv Const. \tag{5.23}$$

Next, we will study condition **(2)**″. Notice that

$$\frac{\partial J}{\partial x^i} = \frac{\partial J}{\partial F_M^L} \frac{\partial F_M^L}{\partial x^i}.$$

Using the derivative of $J$ (which we have shown above), we have that

$$\frac{\partial J}{\partial F_{M|\tilde{F}}^L} = \det(\tilde{F})(\tilde{F}^{-1})_M^L.$$

Then, **(2)**″ is satisfied if and only if

$$\frac{\partial F_M^L}{\partial x^i} = 0, \quad \forall i, L, M. \tag{5.24}$$

Observe that

$$\frac{\partial F_M^L}{\partial x_{|j_{X,Y}^1}^k} = \frac{\partial F_M^L \circ (x^i, y^j, y_i^j)^{-1}}{\partial X_{|(\tilde{X}, \tilde{Y}, \tilde{F})}^K}$$

$$= \frac{\partial}{\partial X_{(\tilde{X}, \tilde{Y}, \tilde{F})}^K} \left( \frac{\partial X^L \circ (y^j)^{-1}}{\partial X_{|\tilde{Y}}^K} \cdot \tilde{F}_R^K \cdot \left[ \frac{\partial X^R \circ (x^i)^{-1}}{\partial X_{|\tilde{X}}^M} \right]^{-1} \right).$$

i.e.,

$$\frac{\partial F_M^L}{\partial x_{|j_{X,Y}^1}^k} = 0,$$

if and only if

$$\frac{\partial}{\partial X_{|\tilde{X}}^K} \left( \frac{\partial X^M \circ (x^i)^{-1}}{\partial X_{|\tilde{X}}^I} \right) = 0.$$

So, **(2)**″ is tantamount to,

$$\frac{\partial X^M}{\partial x^i} \equiv \text{Const}, \quad \forall i, M.$$

This fact implies that,

$$e\left(X^M\right) \equiv \text{Const}, \quad \forall m,$$

i.e.,

$$e = \mu^I \frac{\partial}{\partial X^I}, \quad \mu^I \equiv \text{Const}.$$

Notice that, by using Eq. (5.23), this implies, indeed, that the canonical basis (and hence the reference configuration of $\mathcal{B}$) is a (global) system of homogeneous coordinates on $\mathcal{B}$. So, we extract the following conclusions:

(a) $\mathcal{B}$ is (locally) homogeneous if and only if the vector field $e$ is constant with respect to the canonical basis of $\mathbb{R}^3$.

(b) The homogeneity of $\mathcal{B}$ implies that the reference coordinates are homogeneous coordinates.

(c) $\mathcal{B}$ is locally homogeneous if and only if $\mathcal{B}$ is globally homogeneous.

# Chapter 6

# Characteristic Distributions and Material Bodies

From the existence of structures of simple bodies $\mathcal{B}$ in which the material groupoid $\Omega(\mathcal{B})$ is not a Lie subgroupoid of the groupoid of 1-jets $\Pi^1(\mathcal{B}, \mathcal{B})$ arises the need to develop more "*differentiable tools*". More generally, we will start studying the case of a general subgroupoid $\overline{\Gamma}$ of a Lie groupoid $\Gamma$ to get results which may be applied to material bodies as well as other interesting examples.

## 6.1    Characteristic Distribution

As we have said, sometimes it could be necessary to work with a groupoid which does not have a structure of Lie groupoid. In fact, the constitutive theory of continuum mechanics is an example (see Section 6.2). In this case, the set of material isomorphisms has the structure of subgroupoid of a particular Lie groupoid: the 1-jets groupoid on a manifold. However, this groupoid is not necessarily a Lie subgroupoid of the 1-jets groupoid. This will be discussed in the next chapter in some detail.

In this chapter, we will work with a general subgroupoid of a given Lie groupoid.

Let $\Gamma \rightrightarrows M$ be a Lie groupoid and $\overline{\Gamma}$ be a subgroupoid of $\Gamma$ (not necessarily a Lie subgroupoid of $\Gamma$) over the same manifold $M$. We will denote by $\overline{\alpha}$, $\overline{\beta}$, $\overline{\epsilon}$ and $\overline{i}$ the restrictions of the structure maps $\alpha$, $\beta$, $\epsilon$ and $i$ of $\Gamma$ to $\overline{\Gamma}$ (see the diagram below).

119

where $j$ is the inclusion map. Now, we can construct a distribution $A\overline{\Gamma}^T$ over the manifold $\Gamma$ in the following way:

$$g \in \Gamma \mapsto A\overline{\Gamma}^T_g \le T_g\Gamma,$$

such that $A\overline{\Gamma}^T_g$ is the fibre of $A\overline{\Gamma}^T$ at $g$ and it is generated by the (local) left-invariant vector fields $\Theta \in \mathfrak{X}_{\text{loc}}(\Gamma)$ whose flow at the identities is totally contained in $\overline{\Gamma}$, i.e.,

(i) $\Theta$ is tangent to the $\beta$-fibres,

$$\Theta(g) \in T_g\beta^{-1}(\beta(g)),$$

for all $g$ in the domain of $\Theta$;

(ii) $\Theta$ is invariant by left translations,

$$\Theta(g) = T_{\epsilon(\alpha(g))}L_g(\Theta(\epsilon(\alpha(g)))),$$

for all $g$ in the domain of $\Theta$;

(iii) the (local) flow $\varphi^\Theta_t$ of $\Theta$ satisfies

$$\varphi^\Theta_t(\epsilon(x)) \in \overline{\Gamma},$$

for all $x \in M$.

Notice that, for each $g \in \Gamma$, the zero vector $0_g \in T_g\Gamma$ is contained in the fibre of the distribution at $g$, namely $A\overline{\Gamma}^T_g$ (we remit to the last section for non-trivial examples). On the other hand, it is easy to prove that a vector field $\Theta$ satisfies conditions (i) and (ii) if and only if its local flow $\varphi^\Theta_t$ is left-invariant or, equivalently,

$$L_g \circ \varphi^\Theta_t = \varphi^\Theta_t \circ L_g, \quad \forall g, t.$$

Then, taking into account that all the identities are in $\overline{\Gamma}$ (because it is a subgroupoid of $\Gamma$), condition (iii) is equivalent to the following,

(iii)$'$ The (local) flow $\varphi^\Theta_t$ of $\Theta$ at $\overline{g}$ is totally contained in $\overline{\Gamma}$, for all $\overline{g} \in \overline{\Gamma}$.

Thus, we are taking the left-invariant vector fields on $\Gamma$ whose integral curves are confined inside or outside $\overline{\Gamma}$. It is also remarkable that, by definition, this distribution is differentiable. Remember that a distribution is differentiable (see Appendix A) if for any point $x$ and for any vector $v_x$ of the distribution at $x$ there exists a (local) vector field $\Theta$ tangent to the distribution such that,

$$\Theta(x) = v_x.$$

The distribution $A\overline{\Gamma}^T$ is called the *characteristic distribution of* $\overline{\Gamma}$.

For the sake of simplicity, we will denote the family of the vector fields which satisfy conditions (i), (ii) and (iii) by $\mathcal{C}$. The local vector fields of $\mathcal{C}$ will be called *admissible vector fields*.

**Remark 6.1.1.** Our construction of the characteristic distribution associated to a subgroupoid $\overline{\Gamma}$ of a Lie groupoid $\Gamma$ can be seen as a generalization of the construction of the associated Lie algebroid to a given Lie groupoid (see Chapter 4).

The structure of groupoid permits us to construct two more new objects associated to the distribution $A\overline{\Gamma}^T$. The first one is a smooth distribution over the base $M$ denoted by $A\overline{\Gamma}^\sharp$. The second one is a "differentiable" correspondence $A\overline{\Gamma}$ which associates to any point $x$ of $M$ a vector subspace of $T_{\epsilon(x)}\Gamma$. Both constructions are characterized by the commutativity of the following diagram:

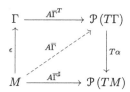

where $\mathcal{P}(E)$ defines the power set of $E$. Therefore, for each $x \in M$, the fibres satisfy that

$$A\overline{\Gamma}_x = A\overline{\Gamma}^T_{\epsilon(x)},$$

$$A\overline{\Gamma}^\sharp_x = T_{\epsilon(x)}\alpha\left(A\overline{\Gamma}_x\right).$$

The distribution $A\overline{\Gamma}^\sharp$ is called *base-characteristic distribution of* $\overline{\Gamma}$. It is remarkable that all the distributions introduced are not, necessarily, regular.

Notice that, taking into account that $A\overline{\Gamma}^T$ is locally generated by left-invariant vector field, we have that for each $g \in \Gamma$,

$$A\overline{\Gamma}^T_g = T_{\epsilon(\alpha(g))}L_g(A\overline{\Gamma}^T_{\epsilon(\alpha(g))}),$$

i.e., the characteristic distribution is *left-invariant*. In particular, the characteristic distribution and the base-characteristic distribution are characterized by the differentiable correspondence $A\overline{\Gamma}$ by using that

$$A\overline{\Gamma}^T_g = T_{\epsilon(\alpha(g))}L_g\left(A\overline{\Gamma}_{\alpha(g)}\right).$$

We could have used Grassmannian manifolds instead of power sets in the above diagram for the distributions but we prefer power sets because of the simplicity.

To summarize, associated to $\overline{\Gamma}$ we have three differentiable objects $A\overline{\Gamma}$, $A\overline{\Gamma}^T$ and $A\overline{\Gamma}^\sharp$. Now, we will study how these objects endow $\overline{\Gamma}$ with a sort of "differentiable" structure. Consider a left-invariant vector field $\Theta$ on $\Gamma$ whose (local) flow $\varphi_t^\Theta$ at the identities is contained in $\overline{\Gamma}$. We want to prove that the characteristic distribution $A\overline{\Gamma}^T$ is invariant by the flow $\varphi_t^\Theta$, i.e., for all $g \in \Gamma$ and $t$ in the domain of $\varphi_g^\Theta$ we have

$$T_g\varphi_t^\Theta(A\overline{\Gamma}^T_g) = A\overline{\Gamma}^T_{\varphi_t^\Theta(g)}. \tag{6.1}$$

Indeed, let $v_g = \Xi(g) \in A\overline{\Gamma}^T_g$ with $\Xi \in \mathcal{C}$. Then,

$$T_g\varphi_t^\Theta(v_g) = T_g\varphi_t^\Theta(\Xi(g))$$

$$= \frac{\partial}{\partial s_{|0}}\left(\varphi_t^\Theta \circ \varphi_s^\Xi(g)\right),$$

where $\varphi_s^\Xi$ is the flow of $\Xi$.

Consider the (local) vector field $\Upsilon$ on $\Gamma$ given by the pullback of $\Xi$ by $\varphi_t^\Theta$, i.e., for each $h$

$$\Upsilon(h) = \{(\varphi_t^\Theta)^*\Xi\}(h) = T_{\varphi_{-t}^\Theta(h)}\varphi_t^\Theta\left(\Xi\left(\varphi_{-t}^\Theta(h)\right)\right).$$

Then, the flow of $\Upsilon$ is $\varphi_t^\Theta \circ \varphi_s^\Xi \circ \varphi_{-t}^\Theta$ and, therefore, $\Upsilon$ is an admissible vector field. Furthermore,

$$T_g\varphi_t^\Theta(v_g) = \Upsilon\left(\varphi_t^\Theta(g)\right).$$

Hence, $T_g \varphi_t^\Theta \left( A\overline{\Gamma}_g^T \right) \subseteq A\overline{\Gamma}_{\varphi_t^\Theta(g)}^T$. We can prove the converse in an analogous way.

Thus, the characteristic distribution $A\overline{\Gamma}^T$ is locally generated by a family of vector fields $\mathcal{C}$, and it is invariant with respect this family. Remember now the celebrated Stefan–Sussmann's theorem (Theorem A.0.17) which deals with the integrability of singular distributions (see Appendix A for a detailed presentation of this result).

**Theorem 6.1.2 (Stefan–Sussmann).** *Let $D$ be a smooth singular distribution on a smooth manifold $M$. Then the following three conditions are equivalent:*

(a) *$D$ is integrable.*

(b) *$D$ is generated by a family $C$ of smooth vector fields, and is invariant with respect to $C$.*

(c) *$D$ is the tangent distribution $D^{\mathcal{F}}$ of a smooth singular foliation $\mathcal{F}$.*

There is still another theorem to deal with the integrability of generalized distributions which could be confused with the Stefan–Sussmann's theorem, *Hermann's theorem* (Theorem A.0.22), which states that any locally finitely generated differentiable involutive distribution on a manifold is integrable.

So, the distribution $A\overline{\Gamma}^T$ is the tangent distribution of a smooth singular foliation $\overline{\mathcal{F}}$. The leaf at a point $g \in \Gamma$ is denoted by $\overline{\mathcal{F}}(g)$. The collection of the leaves of $\overline{\mathcal{F}}$ at points of $\overline{\Gamma}$ is called the *characteristic foliation of $\overline{\Gamma}$*. Note that the leaves of the characteristic foliation covers $\overline{\Gamma}$ but it is not exactly a foliation of $\overline{\Gamma}$ (because $\overline{\Gamma}$ is not a manifold).

The following assertions can be easily proved:

(i) For each $g \in \Gamma$,

$$\overline{\mathcal{F}}(g) \subseteq \Gamma^{\beta(g)}.$$

Indeed, if $g \in \overline{\Gamma}$, then

$$\overline{\mathcal{F}}(g) \subseteq \overline{\Gamma}^{\beta(g)}.$$

(ii) For each $g, h \in \Gamma$ such that $\alpha(g) = \beta(h)$, we have

$$\overline{\mathcal{F}}(g \cdot h) = g \cdot \overline{\mathcal{F}}(h).$$

The property (ii) is proved by arguments of maximality. On the other hand, the property (i) can be proved by checking the charts of the leaves given in the proof of the Stefan–Sussmann's theorem (see the proof of Theorem A.0.17). It is remarkable that property (i) means that each leaf of the foliation $\overline{\mathcal{F}}$ which integrates $A\overline{\Gamma}^T$ is contained in just one $\beta$-fibre, i.e., for each $g \in \Gamma$ the leaf $\overline{\mathcal{F}}(g)$ satisfies that

$$\beta(h) = \beta(g),$$

for all $h \in \overline{\mathcal{F}}(g)$. Notice also that, one could expect that $\overline{\mathcal{F}}(g) = \overline{\Gamma}^{\beta(g)}$ but this is not true in general (see examples in Section 6.2).

So, we have proved the following result.

**Theorem 6.1.3.** *Let* $\Gamma \rightrightarrows M$ *be a Lie groupoid and* $\overline{\Gamma}$ *be a subgroupoid of* $\Gamma$ *(not necessarily a Lie groupoid) over* $M$. *Then, there exists a foliation* $\overline{\mathcal{F}}$ *of* $\Gamma$ *such that* $\overline{\Gamma}$ *is a union of leaves of* $\overline{\mathcal{F}}$.

In this way, without assuming that $\overline{\Gamma}$ is a manifold, we prove that $\overline{\Gamma}$ is union of leaves of a foliation of $\Gamma$. This gives us some kind of "differentiable" structure over $\overline{\Gamma}$.

Let us consider a (local) left-invariant vector field $\Theta \in \mathcal{C}$. Then, the flow of $\Theta$ restricts to the fibres, i.e., $\Theta$ is a left-invariant vector field in $\Gamma$ such that,

$$\Theta_{|\overline{\mathcal{F}}(g)} \in \mathfrak{X}\left(\overline{\mathcal{F}}(g)\right), \tag{6.2}$$

for all $g$ in the domain of $\Theta$. Reciprocally, left-invariant vector fields satisfying Eq. (6.2) are admissible vector fields.

It is important to note that we are working with the case in which $\overline{\Gamma}$ is a subgroupoid of $\Gamma$ over the same manifold. In fact, we could do not have to impose it.

**Corollary 6.1.4.** *Let* $\Gamma \rightrightarrows M$ *be a Lie groupoid and* $\overline{\Gamma}$ *be a subgroupoid of* $\Gamma$ *(not necessarily a Lie groupoid). Then, there exists a maximal foliation* $\overline{\mathcal{F}}$ *of* $\Gamma$ *such that* $\overline{\Gamma}$ *is a union of leaves of* $\overline{\mathcal{F}}$.

**Proof.** Denote by $N$ to the subset of $M$ given by $\alpha(\overline{\Gamma}) = \beta(\overline{\Gamma})$. Then, $N$ is not necessarily a submanifold of $M$.

Let us now define the following subgroupoid $\overline{\Gamma}_M$ over $M$:

$$\overline{\Gamma}_M = \Gamma_{M-N} \sqcup \overline{\Gamma},$$

where $\Gamma_{M-N}$ is the set of all the elements of $\Gamma$ from points of $M - N$ to points of $M - N$, with $M - N$ as the collection of points of $M$

outside $N$. Observe that all the identities at points of $M$ are in $\overline{\Gamma}_M$. It is also remarkable that $\overline{\Gamma}_M$ is now a subgroupoid of $\Gamma$ over the same manifold $M$. Then, we apply Theorem 6.1.3 to $\overline{\Gamma}_M$ to get the characteristic foliation $\overline{\mathcal{F}}$ of $\Gamma$ such that $\overline{\Gamma}_M$ is a union of leaves of $\overline{\mathcal{F}}$.

Let $x \in M$ be a point which is not at $N$. Then,

$$\overline{\mathcal{F}}(\epsilon(x)) \subseteq \overline{\Gamma}_M^x \subseteq \Gamma_{M-N}.$$

Hence, the foliation $\overline{\mathcal{F}}$ of $\Gamma$ satisfies that $\overline{\Gamma}$ is a union of leaves of $\overline{\mathcal{F}}$. □

Let $\overline{\Gamma}_M^2$ be another subgroupoid of $\Gamma$ over $M$ extending $\overline{\Gamma}$ such that there are not elements in $\overline{\Gamma}_M^2$ based at points of $N$ which are not in $\overline{\Gamma}$, i.e., for all $g \in \overline{\Gamma}_M^2$ such that $\alpha(g) \in N$ it satisfies that $g \in \overline{\Gamma}$. Equivalently, there are not elements $g \in \overline{\Gamma}_M^2$ with $\beta(g) \in N$ outside $\overline{\Gamma}$. Notice that this is a natural imposition if we want that the characteristic foliation restricts to $\overline{\Gamma}$.

Then, there exists a set $A$ of elements of $\Gamma$ from points of $M - N$ to points of $M - N$ containing the identities satisfying that

(i) $\overline{\Gamma}_M^2 = A \sqcup \overline{\Gamma}$;
(ii) $\epsilon(M - N) \subseteq A \subseteq \Gamma_{M-N}$.

Denote by $\mathcal{C}_M^A$ and $\mathcal{C}_M$ to the admissible vector fields associated to $\overline{\Gamma}_M^2$ and $\overline{\Gamma}_M$, respectively. Then, by using **(i)** and **(ii)** we have that

$$\mathcal{C}_M^A \subseteq \mathcal{C}_M.$$

Therefore, all the leaves of the characteristic foliation associated to $\overline{\Gamma}_M^2$ are contained in the leaves of the characteristic foliation associated to $\overline{\Gamma}_M$. Then, the leaves associated to $\overline{\Gamma}_M$ are maximal and this is the reason because we chose the extension $\overline{\Gamma}_M$.

One could think that, maybe, the resulting leaves inside $N$ do not depend on the choice of the subset $A$ but this is not true (see example below).

**Example 6.1.5.** Consider $M = \mathbb{R}^2$, $\Gamma = \mathbb{R}^2 \times \mathbb{R}^2$ and $\overline{\Gamma} = \overline{B}_1(0) \times \overline{B}_1(0)$, where $\overline{B}_1(0)$ is the closed ball of centre 0 and radius 1 (what implies that $N = \alpha\left(\overline{\Gamma}\right) = \beta\left(\overline{\Gamma}\right) = \overline{B}_1\left(0\right))$. Then, consider

$$\overline{\Gamma}^2_M = A \sqcup \overline{\Gamma},$$

with the set $A$ defined as follows:

$$A = \epsilon\left(\overline{B}_1\left(0\right)^c\right) = \{(x,x) : x \notin \overline{B}_1\left(0\right)\},$$

with $\overline{B}_1\left(0\right)^c$ the complementary set of $\overline{B}_1\left(0\right)$ in $\mathbb{R}^2$.

So, for each point outside $N$ we only add the identity at the point. Then, in this case, an admissible vector field has to be zero at points outside $N$. By continuity, the admissible vector fields are also necessarily zero at the boundary of $\overline{B}_1\left(0\right)$, i.e., at the sphere $S\left(1\right)$ of centre 0 and radius 1 in $\mathbb{R}^2$. On the other hand, inside the open ball $B_1\left(0\right)$ any local vector $\Theta$ field induce an admissible vector field as follows:

$$\Theta^l\left(x,y\right) = \left(\Theta\left(x\right),0\right) \tag{6.3}$$

for all $x,y$ in the domain of $\Theta$. Therefore, the characteristic foliation $\overline{\mathcal{F}}_A$ of $\overline{\Gamma}^2_M$ is given by

(i) For all point $x \in S\left(1\right)$,

$$\overline{\mathcal{F}}_A\left(x,y\right) = \{x\} \times \{y\},$$

for any $y \in \mathbb{R}^2$.

(ii) For all point $x \in B_1\left(0\right)$,

$$\mathcal{F}_A\left(x,y\right) = B_1\left(0\right) \times \{y\},$$

for any $y \in \mathbb{R}^2$.

On the other hand, let us take the extension considered in the proof of Corollary 6.1.4, i.e.,

$$\overline{\Gamma}_M = \left[\overline{B}_1\left(0\right)^c \times \overline{B}_1\left(0\right)^c\right] \sqcup \left[\overline{B}_1\left(0\right) \times \overline{B}_1\left(0\right)\right].$$

Then, taking into account that $\overline{B}_1\left(0\right)^c$ is an open subset of $\mathbb{R}^2$, any (local) vector field at $\overline{B}_1\left(0\right)^c$ induces an admissible vector field by Eq. (6.3). Analogously, any (local) vector field at $B_1\left(0\right)$ induces an admissible vector field for $\overline{\Gamma}_M$.

Let $\Theta$ be a (local) vector field with domain $U$ such that $U \cap S(1) \neq \emptyset$. It is easy to see that if $\Theta$ is tangent to the sphere $S\left(1\right)$

then, $\Theta$ induces an admissible vector field for $\overline{\Gamma}_M$ by using Eq. (6.3). So, the characteristic distribution $A\overline{\Gamma}_M^T$ has dimension larger than 1. On the other hand, there are vector fields which are not admissible vector fields, for instance,

$$\Theta\left(x,y\right) = \left(\frac{\partial}{\partial r^1}, 0\right),$$

where $\left(r^i\right)$ is the (global) canonical system of coordinates of $\mathbb{R}^2$, is not an admissible vector field. Therefore, necessarily,

$$\dim\left(\left(A\overline{\Gamma}_M\right)^T_{(x,y)}\right) = 1.$$

for any $x \in S\left(1\right)$ and $y \in \mathbb{R}^2$. Thus, we have proved that the characteristic foliation $\overline{\mathcal{F}}$ associated to $\overline{\Gamma}_M$ is given by

**(I)** For all point $x \in S\left(1\right)$,

$$\overline{\mathcal{F}}\left(x,y\right) = S\left(1\right) \times \{y\},$$

for any $y \in \mathbb{R}^2$.

**(II)** For all point $x \in B_1\left(0\right)$,

$$\overline{\mathcal{F}}\left(x,y\right) = B_1\left(0\right) \times \{y\},$$

for any $y \in \mathbb{R}^2$.

In this way, these extensions ($\overline{\Gamma}_M$ and $\overline{\Gamma}_M^2$) generate strictly different characteristic leaves inside $\overline{\Gamma}$.

For a general subgroupoid $\overline{\Gamma}$ of a Lie groupoid $\Gamma$, we will call *characteristic distribution of* $\overline{\Gamma}$ to the characteristic distribution of $\overline{\Gamma}_M$ and it will be denoted by $A\overline{\Gamma}^T$. The collection of the leaves of $\overline{\mathcal{F}}$ at points of $\overline{\Gamma}$ is again called the *characteristic foliation of* $\overline{\Gamma}$.

Obviously, from the construction of the characteristic distribution, we obtain some condition of maximality.

**Corollary 6.1.6.** *Let* $\overline{\mathcal{G}}$ *be a left invariant foliation of* $\Gamma$ *such that* $\overline{\Gamma}$ *is a union of leaves of* $\overline{\mathcal{G}}$. *Then, the characteristic foliation* $\overline{\mathcal{F}}$ *is coarser that* $\overline{\mathcal{G}}$, *i.e.,*

$$\overline{\mathcal{G}}\left(g\right) \subseteq \overline{\mathcal{F}}\left(g\right), \quad \forall g \in \Gamma. \tag{6.4}$$

**Proof.** Taking into account Corollary 6.1.4 we may assume that $\overline{\Gamma}$ is a subgroupoid of $\Gamma$ over the same manifold $M$. Let $\mathcal{D}$ be the family of (local) vector fields tangent to the foliation $\overline{\mathcal{G}}$. Then, the left-invariance of $\overline{\mathcal{G}}$ implies that any vector field $\Theta \in \mathcal{D}$ is tangent to the $\beta$-fibres. So, we may define a new left-invariant vector field $\Theta_L$ such that for each $g$,

$$\Theta_L(g) = T_{\epsilon(\alpha(g))} L_g \left( \Theta \left( \epsilon \left( \alpha \left( g \right) \right) \right) \right).$$

Denote the family of left-invariant vector fields induced by the vector field in $\mathcal{D}$ by $\mathcal{D}_L$. Then, $\mathcal{D}_L$ generates the tangent distribution to $\overline{\mathcal{G}}$. In fact, for each $x \in M$, $T_{\epsilon(x)} \overline{\mathcal{G}} \left( \epsilon \left( x \right) \right)$ is obviously generated by the evaluation of the vector fields of $\mathcal{D}_L$ at the identity $\epsilon \left( x \right)$ (the evaluation of the vector fields of $\mathcal{D}_L$ at the identity $\epsilon \left( x \right)$ results exactly in the vector fields of $\mathcal{D}$ at the identity $\epsilon \left( x \right)$). Thus, the left-invariance of $\overline{\mathcal{G}}$ proves that $\mathcal{D}_L$ generates the tangent distribution to $\overline{\mathcal{G}}$.

Finally, using that $\overline{\Gamma}$ is a union of leaves of $\overline{\mathcal{G}}$ we have that $\mathcal{D}_L \subseteq \mathcal{C}$ and, therefore, Eq. (6.4) is satisfied. $\qquad\square$

Particularly, $\overline{\Gamma}^x$ *is a submanifold of* $\Gamma$ *for all* $x \in M$ *if and only if* $\overline{\Gamma}^x = \overline{\mathcal{F}}\left( \epsilon \left( x \right) \right)$ *for all* $x \in M$.

Notice that, in an analogous way to Theorem 6.1.3, we can prove that the base-characteristic distribution $A\overline{\Gamma}^\sharp$ is also integrable. Thus, we will denote the resulting foliation which integrates the base-characteristic distribution over the base $M$ by $\mathcal{F}$. For each point $x \in M$, the leaf of $\mathcal{F}$ through $x$ will be denoted by $\mathcal{F}(x)$. $\mathcal{F}$ will be called the *base-characteristic foliation of* $\overline{\Gamma}$.

Let us apply these results to a particular example. Let $M$ be a manifold and $M \times M$ the pair groupoid (Example 3.0.6). Then, any transitive subgroupoid of $M$ is the pair groupoid $N \times N$ of a subset $N \subseteq M$. Then, using Corollary 6.1.4 we have the following result.

**Theorem 6.1.7.** *Let* $M$ *be a manifold and* $N$ *be a subset of* $M$. *Then, there exists a maximal foliation* $\mathcal{F}$ *of* $M$ *such that* $N$ *is union of leaves.*

**Proof.** Let $N \times N \rightrightarrows N$ be the transitive pair groupoid of $N$. We will consider the subgroupoid

$$(N \times N)_M = [(M - N) \times (M - N)] \sqcup [N \times N] \rightrightarrows M,$$

of $M \times M \rightrightarrows M$. So, we may consider $\overline{\mathcal{F}}$ and $\mathcal{F}$ its characteristic foliation and base-characteristic foliation respectively.

Then, for each $x \in N$ we have that

$$\overline{\mathcal{F}}(x, x) \subseteq N \times \{x\}.$$

In fact, it satisfies that

$$\overline{\mathcal{F}}(x, x) = \mathcal{F}(x) \times \{x\}. \tag{6.5}$$

Hence, $N$ is the union of the leaves of the base-characteristic foliation at points of $N$ and we already have our foliation.  □

Let us now explain the condition of maximality of the foliation. Let $\mathcal{G}$ be another foliation of $M$ such that $N$ is union of leaves. Then, for each $(x, y) \in M \times M$ we may define

$$\overline{\mathcal{G}}(x, y) = \mathcal{G}(x) \times \{y\}.$$

Hence, the family $\overline{\mathcal{G}} = \{\overline{\mathcal{G}}(x, y)\}_{(x,y) \in M \times M}$ defines a left invariant foliation of $M \times M$ such that $[(M - N) \times (M - N)]$ is union of leaves. Thus, the maximality condition of the characteristic foliation (Corollary 6.1.6) implies that $\mathcal{G} \subset \mathcal{F}$, i.e., there is no another coarser foliation of $M$ which divides $N$ into union of leaves.

Notice that the maximal foliation given in Theorem 6.1.7 permits us to endow $N$ with differential structure which generalizes the structure of manifold. Indeed, *N is a submanifold of M if and only if N consists of just one leaf of the foliation.*

Let $\Theta$ be an admissible vector field of the subgroupoid $[(M - N) \times (M - N)] \sqcup [N \times N] \rightrightarrows M$. Then, the projection

$$\theta = T\alpha \circ \Theta \circ \epsilon,$$

on $M$ is a vector field on $M$ such that its flow at point of $N$ is confined in $N$. Conversely, any vector fields $\theta$ whose flows at point of $N$ is inside $N$ may be lifted to an admissible vector field $\Theta$ by imposing that

$$\Theta(x, y) = (\theta(x), 0) \in T_x M \times T_y M, \quad \forall x, y \in M. \tag{6.6}$$

Thus, the foliation given in the Theorem 6.1.7 can be described by the vector fields on $M$ whose flow at points of $N$ is contained in $N$.

**Example 6.1.8.** Let $\sim$ be an equivalence relation on a manifold $M$, i.e., a binary relation that is reflexive, symmetric and transitive.

Then, define the subset $\mathcal{O}$ of $M \times M$ given by

$$\mathcal{O} := \{(x, y) : x \sim y\}. \tag{6.7}$$

Then, $\mathcal{O}$ is a subgroupoid of $M \times M$ over $M$. In fact, this is equivalent to the properties reflexive, symmetric and transitive. For each $x \in M$, we denote by $\mathcal{O}_x$ to the orbit around $x$,

$$\mathcal{O}_x := \{y : x \sim y\}.$$

Notice that the orbits divide $M$ into a disjoint union of subsets. However, these are not (necessarily) submanifolds.

On the other hand, the base-characteristic foliation gives us a foliation $\mathcal{F}$ of $M$ such that

$$\mathcal{F}(x) \subseteq \mathcal{O}_x, \quad \forall x \in M.$$

This foliation is maximal in the sense that there is no any other coarser foliation of $M$ whose leaves are contained in the orbits (see Theorem 6.1.13 and Corollary 6.1.14).

Another example give rise to the so-called *material distributions*. This example will be presented in the next section.

**Remark 6.1.9.** We can construct another distribution $\mathcal{D}$ on $\overline{\Gamma}$ generated by the (local) vector fields whose flows are confined inside or outside $\overline{\Gamma}$. So, we will obtain a foliation $\overline{\mathcal{G}}$ of $\Gamma$ such that $\overline{\Gamma}$ is covered by some of the leaves.

We could expect that the leaves at the identities $\overline{\mathcal{G}}(\epsilon(x))$ are subgroupoids of $\Gamma$. However, this is not necessarily true. Because of this fact, we work with $A\overline{\Gamma}^T$ instead of $\mathcal{D}$ (see Theorem 6.1.13).

Next, we will prove that the leaves of $\mathcal{F}$ have even more geometric structure. In fact, we will find a Lie groupoid structure over each leaf of $\mathcal{F}$. To do this, we will prove the following technical proposition.

**Proposition 6.1.10.** *Let $\Gamma \rightrightarrows M$ be a Lie groupoid and $\overline{\Gamma}$ be a subgroupoid of $\Gamma$ with $\overline{\mathcal{F}}$ and $\mathcal{F}$ the characteristic foliation and the base-characteristic foliation of $\overline{\Gamma}$, respectively. Then, for all $x \in M$,* *the mapping*

$$\alpha_{|\overline{\mathcal{F}}(\epsilon(x))} : \overline{\mathcal{F}}(\epsilon(x)) \to \mathcal{F}(x),$$

*is a surjective submersion.*

**Proof.**  First, let us notice that

$$x \in \alpha\left(\overline{\mathcal{F}}(\epsilon(x))\right) \cap \mathcal{F}(x) \neq \emptyset.$$

Next, consider a family $\{\Theta^i, \Xi^j\}_{i=1,\dots,r,j=1,\dots,s}$ of left-invariant vector fields in $\mathcal{C}$ such that $\{T_{\epsilon(x)}\alpha\left(\Theta^i\left(\epsilon(x)\right)\right)\}_{i=1,\dots,r}$ is a basis of $A\overline{\Gamma}_x^\sharp$ and $\{\Theta^i\left(\epsilon(x)\right), \Xi^j\left(\epsilon(x)\right)\}_{i=1,\dots,r,j=1,\dots,s}$ is a basis of $A\overline{\Gamma}_{\epsilon(x)}^T$.

Notice that the family $\{T\alpha \circ \Theta^i \circ \epsilon, T\alpha \circ \Xi^j \circ \epsilon\}_{i=1,\dots,r,\ j=1,\dots,s}$ of vector fields on $M$ is tangent to the base-characteristic distribution $A\overline{\Gamma}^\sharp$. So, their flows at $x$ are contained in $\mathcal{F}(x)$.

Furthermore, the map

$$\alpha \circ \varphi_{t_1}^{\Theta^1} \circ \epsilon \circ \cdots \circ \alpha \circ \varphi_{t_r}^{\Theta^r}(\epsilon(x)) = \alpha(\varphi_{t_1}^{\Theta^1} \circ \cdots \circ \varphi_{t_r}^{\Theta^r}(\epsilon(x))),$$

defines a local chart of $\mathcal{F}(x)$ containing $x$, where $\varphi_{t_i}^{\Theta^i}$ is the (local) flow of $\Theta^i$ for each $i$. Following this argument, one can prove that $\alpha\left(\overline{\mathcal{F}}(\epsilon(x))\right)$ is an open subset of $\mathcal{F}(x)$.

Then, $\mathcal{F}(x)$ is the disjoint union of open subsets. Using that $\mathcal{F}(x)$ is connected we have that

$$\alpha\left(\overline{\mathcal{F}}(\epsilon(x))\right) = \mathcal{F}(x),$$

i.e., $\alpha_{|\overline{\mathcal{F}}(\epsilon(x))}$ is surjective. Hence, $\alpha_{|\overline{\mathcal{F}}(\epsilon(x))}$ is a submersion.  □

Let $x \in M$ and $\Theta \in \mathfrak{X}(\mathcal{F}(x))$. Then, by using local sections of $\alpha_{|\overline{\mathcal{F}}(\epsilon(x))}$, we can extend (locally) $\Theta$ to a (left-invariant) vector field on $\overline{\mathcal{F}}(\epsilon(x))$. In this way, $\Theta$ is a local vector field tangent to the base-characteristic distribution if and only if it satisfies that

$$\Theta_{|\mathcal{F}(x)} \in \mathfrak{X}(\mathcal{F}(x)), \qquad (6.8)$$

for all $x$ in the domain of $\Theta$.

As a corollary, we have the following interesting result.

**Corollary 6.1.11.** *Let* $\Gamma \rightrightarrows M$ *be a Lie groupoid and* $\overline{\Gamma}$ *be a subgroupoid of* $\Gamma$. *Then, the manifolds* $\overline{\mathcal{F}}(\epsilon(x)) \cap \alpha^{-1}(x)$ *are Lie subgroups of* $\Gamma_x^x$ *for all* $x \in M$.

**Proof.** Let $h, g \in \overline{\mathcal{F}}(\epsilon(x)) \cap \alpha^{-1}(x)$. Then,

$$\overline{\mathcal{F}}(h \cdot g) = h \cdot \overline{\mathcal{F}}(g) = h \cdot \overline{\mathcal{F}}(\epsilon(x)) = \overline{\mathcal{F}}(h) = \overline{\mathcal{F}}(\epsilon(x)). \qquad \square$$

Another interesting consequence is that we can improve Corollary 6.1.6.

**Corollary 6.1.12.** *Let $\overline{\mathcal{G}}$ be a foliation of $\Gamma$ such that $\overline{\Gamma}$ is a union of leaves of $\overline{\mathcal{G}}$ and*

$$\overline{\mathcal{G}}(g) \subseteq \Gamma^{\beta(g)}, \quad \forall g \in \Gamma.$$

*Then, the characteristic foliation $\overline{\mathcal{F}}$ is coarser that $\overline{\mathcal{G}}$, i.e.,*

$$\overline{\mathcal{G}}(g) \subseteq \overline{\mathcal{F}}(g), \quad \forall g \in \Gamma.$$

**Proof.** Let us consider $\mathcal{D}$ as the family of (local) vector fields tangent to the foliation $\overline{\mathcal{G}}$. Fix $g \in \Gamma$ and $v_g \in T_g \overline{\mathcal{G}}(g)$. We may assume that there exists $\Theta \in \mathcal{D}$ such that

$$\Theta(g) = v_g. \tag{6.9}$$

By using Proposition 6.1.10, we may have a local section $\sigma_g$ of $\alpha_{|g \cdot \overline{\mathcal{F}}(\epsilon(\alpha(g)))} : g \cdot \overline{\mathcal{F}}(\epsilon(\alpha(g))) \to \mathcal{F}(\alpha(g))$ with $\sigma_g(\alpha(g)) = g$. So, we will define the following (local) left-invariant vector field $\Upsilon^{\sigma_g}$ on $g \cdot \overline{\mathcal{F}}(\epsilon(\alpha(g)))$ characterized by

$$\Upsilon^{\sigma_g}(\epsilon(y)) = T_{\sigma_g(y)} L_{\sigma_g(y)^{-1}}(\Theta(\sigma_g(y))). \tag{6.10}$$

Thus, the flow of $\Upsilon^{\sigma_g}$ is given by

$$\varphi_t^{\Upsilon^{\sigma_g}}(h) = h \cdot (\sigma_g(\alpha(h))^{-1}) \cdot \varphi_t^{\Theta}(\sigma_g(\alpha(h))).$$

Hence, $\Upsilon^{\sigma_g}$ generates an admissible vector field. Furthermore,

$$\Upsilon^{\sigma_g}(g) = \Theta(g) = v_g,$$

i.e., $v_g \in A\overline{\Gamma}_g^T$. $\qquad \square$

Notice that, taking into account this result, we may "relax" the conditions of the family of admissible vector fields. In fact, the characteristic distribution is generated by the (local) vector fields $\Theta \in \mathfrak{X}_{\text{loc}}(\Gamma)$ such that

(i) $\Theta$ is tangent to the $\beta$-fibres,

$$\Theta\left(g\right) \in T_g \Gamma^{\beta(g)};$$

for all $g$ in the domain of $\Theta$;

(ii) the (local) flow $\varphi_t^\Theta$ of $\Theta$ satisfies

$$\varphi_t^\Theta\left(\overline{g}\right) \in \overline{\Gamma},$$

for all $\overline{g} \in \overline{\Gamma}$.

Let us now construct an algebraic structure of a groupoid over the leaves of $\mathcal{F}$. We will consider the groupoid $\overline{\Gamma}\left(\mathcal{F}\left(x\right)\right)$ generated by $\overline{\mathcal{F}}\left(\epsilon\left(x\right)\right)$ by imposing that for all $\overline{g}, \overline{h} \in \overline{\mathcal{F}}\left(\epsilon\left(x\right)\right)$,

$$\overline{g}, \overline{g}^{-1}, \overline{h}^{-1} \cdot \overline{g} \in \overline{\Gamma}\left(\mathcal{F}\left(x\right)\right).$$

Notice that

$$\overline{\mathcal{F}}\left(\epsilon\left(x\right)\right) = \overline{\mathcal{F}}\left(\overline{h}\right) = \overline{h} \cdot \overline{\mathcal{F}}\left(\epsilon\left(\alpha\left(\overline{h}\right)\right)\right).$$

Therefore,

$$\overline{\mathcal{F}}(\overline{h}^{-1}) = \overline{h}^{-1} \cdot \overline{\mathcal{F}}\left(\epsilon\left(x\right)\right) = \overline{\mathcal{F}}\left(\epsilon\left(\alpha\left(\overline{h}\right)\right)\right).$$

On the other hand, let be $\overline{t} \in \overline{\mathcal{F}}\left(\epsilon\left(\alpha\left(\overline{h}\right)\right)\right)$. Then,

$$\overline{\mathcal{F}}\left(\overline{h} \cdot \overline{t}\right) = \overline{h} \cdot \overline{\mathcal{F}}\left(\overline{t}\right) = \overline{h} \cdot \overline{\mathcal{F}}\left(\epsilon\left(\alpha\left(\overline{h}\right)\right)\right) = \overline{\mathcal{F}}\left(\epsilon\left(x\right)\right),$$

i.e., $\overline{h} \cdot \overline{t} \in \overline{\mathcal{F}}\left(\epsilon\left(x\right)\right)$ and, hence, $\overline{t}$ can be written as $\overline{h}^{-1} \cdot \overline{g}$ with $\overline{g} \in \overline{\mathcal{F}}(\epsilon(x))$. Thus, we have proved that

$$\overline{\mathcal{F}}\left(\epsilon\left(\alpha\left(\overline{h}\right)\right)\right) \subset \overline{\Gamma}(\mathcal{F}(x)),$$

for all $\overline{h} \in \overline{\mathcal{F}}\left(\epsilon\left(x\right)\right)$. In fact, by following the same argument we have that

$$\overline{\Gamma}\left(\mathcal{F}\left(x\right)\right) = \bigsqcup_{\overline{g} \in \overline{\mathcal{F}}(\epsilon(x))} \overline{\mathcal{F}}(\epsilon(\alpha(\overline{g}))), \tag{6.11}$$

i.e., $\overline{\Gamma}\left(\mathcal{F}\left(x\right)\right)$ can be depicted as a disjoint union of fibres at the identities. Let us now show that $\overline{\Gamma}\left(\mathcal{F}\left(x\right)\right)$ is, in fact, a subgroupoid of $\overline{\Gamma}$. Consider two arbitrary elements $\overline{g}, \overline{h} \in \overline{\Gamma}\left(\mathcal{F}\left(x\right)\right)$ with $\alpha\left(\overline{h}\right) = \beta\left(\overline{g}\right)$. Then, we may assume that we are in one of the following cases:

**(i)** $\overline{g}, \overline{h} \in \overline{\mathcal{F}}(\epsilon(x))$. Then,

$$\overline{\mathcal{F}}(\overline{h} \cdot \overline{g}) = \overline{h} \cdot \overline{\mathcal{F}}(\overline{g}) = \overline{h} \cdot \overline{\mathcal{F}}(\epsilon(x))$$
$$= \overline{\mathcal{F}}(\overline{h}) = \overline{\mathcal{F}}(\epsilon(x)),$$

i.e., $\overline{h} \cdot \overline{g} \in \overline{\mathcal{F}}(\epsilon(x)) \subset \overline{\Gamma}(\mathcal{F}(x))$.

**(ii)** $\overline{g}^{-1}, \overline{h} \in \overline{\mathcal{F}}(\epsilon(x))$. Then,

$$\overline{\mathcal{F}}(\overline{h} \cdot \overline{g}) = \overline{h} \cdot \overline{\mathcal{F}}(\overline{g}) = \overline{h} \cdot \overline{\mathcal{F}}(\epsilon(\beta(\overline{g})))$$
$$= \overline{\mathcal{F}}(\overline{h}) = \overline{\mathcal{F}}(\epsilon(x)).$$

So, $\overline{h} \cdot \overline{g} \in \overline{\mathcal{F}}(\epsilon(x)) \subset \overline{\Gamma}(\mathcal{F}(x))$.

**(iii)** $\overline{g}, \overline{h}^{-1} \in \overline{\mathcal{F}}(\epsilon(x))$. Then,

$$\overline{\mathcal{F}}(\overline{h} \cdot \overline{g}) = \overline{h} \cdot \overline{\mathcal{F}}(\overline{g}) = \overline{h} \cdot \overline{\mathcal{F}}(\epsilon(x))$$
$$= \overline{\mathcal{F}}(\overline{h}) = \overline{\mathcal{F}}(\epsilon(\beta(\overline{h}))).$$

Hence, $\overline{h} \cdot \overline{g} \in \overline{\mathcal{F}}(\epsilon(\beta(\overline{h}))) \subset \overline{\Gamma}(\mathcal{F}(x))$ (see Eq. (6.11)).

It is important to note that $\overline{\Gamma}(\mathcal{F}(x))$ may be equivalently defined as the smallest transitive subgroupoid of $\overline{\Gamma}$ which contains $\overline{\mathcal{F}}(\epsilon(x))$. Observe that the $\beta$-fibre of this groupoid at a point $y \in \mathcal{F}(x)$ is given by $\overline{\mathcal{F}}(\epsilon(y))$. Hence, the $\alpha$-fibre at $y$ is

$$\overline{\mathcal{F}}^{-1}(\epsilon(y)) = i \circ \overline{\mathcal{F}}(\epsilon(y)).$$

Furthermore, the Lie groups $\overline{\mathcal{F}}(\epsilon(y)) \cap \Gamma_y$ are exactly the isotropy groups of $\overline{\Gamma}(\mathcal{F}(x))$. All these results imply the following one (Jiménez *et al.*, 2018).

**Theorem 6.1.13.** *For each $x \in M$, there exists a transitive Lie subgroupoid $\overline{\Gamma}(\mathcal{F}(x))$ of $\Gamma$ with base $\mathcal{F}(x)$.*

**Proof.** Let $g \in \overline{\Gamma}(\mathcal{F}(x))$. Then, by Proposition 6.1.10, the restriction

$$\beta_{|\overline{\mathcal{F}}^{-1}(g)} : \overline{\mathcal{F}}^{-1}(g^{-1}) \to \mathcal{F}(x), \tag{6.12}$$

is a surjective submersion, where $\overline{\mathcal{F}}^{-1}(g^{-1}) = i \circ \overline{\mathcal{F}}(g^{-1})$. Using this fact, we will endow with a differentiable structure to $\overline{\Gamma}(\mathcal{F}(x))$.

Let $g \in \overline{\Gamma}(\mathcal{F}(x))$. Consider $\sigma_g : U \to \overline{\mathcal{F}}^{-1}(g^{-1})$ a (local) section of $\beta_{|\overline{\mathcal{F}}^{-1}(g^{-1})}$ such that $\sigma_g(\beta(g)) = g$.

On the other hand, let $\{\Theta_i\}_{i=1}^r$ be a finite collection of vector fields in $\mathcal{C}$ such that $\{\Theta_i(\epsilon(\alpha(g)))\}_{i=1}^r$ is a basis of $A\overline{\Gamma}^T_{\epsilon(\alpha(g))}$. Then, a local chart $\varphi^\Theta : W \times U \to \Gamma$ over $g$ can be given by

$$\varphi^\Theta(t_1, \ldots, t_r, z) = \sigma_g(z) \cdot [\varphi^{\Theta^r}_{t_r} \circ \cdots \circ \varphi^{\Theta^1}_{t_1}(\epsilon(\alpha(g)))]$$

where $\varphi^{\Theta^i}_t$ is the flow of $\Theta^i$. By using that $\{\Theta^i(\epsilon(\alpha(g)))\}_{i=1}^r$ is a basis of $A\overline{\Gamma}^T_{\epsilon(\alpha(g))}$, we have that $\varphi^\Theta$ is an immersion. Also, it satisfies that

$$\varphi^\Theta(W \times U) \subseteq \overline{\Gamma}.$$

So, these charts give us an atlas over $\overline{\Gamma}(\mathcal{F}(x))$ which induces a Hausdorff second countable topology on $\overline{\Gamma}(\mathcal{F}(x))$ such that $\overline{\Gamma}(\mathcal{F}(x))$ is an immersed submanifold of $\Gamma$. To end the proof we just have to use Eq. (6.12) to prove that the source and the target mappings are submersions. $\qquad\square$

Thus, we have divided the manifold $M$ into leaves $\mathcal{F}(x)$ which have a maximal structure of transitive Lie subgroupoids of $\Gamma$.

**Corollary 6.1.14.** *Let $\mathcal{G}$ be a foliation of $M$ such that for each $x \in M$ there exists a transitive Lie subgroupoid $\Gamma(x)$ of $\Gamma$ over the leaf $\mathcal{G}(x)$ contained in $\overline{\Gamma}$ whose family of $\beta$-fibres defines a foliation on $\Gamma$. Then, the base-characteristic foliation $\mathcal{F}$ is coarser than $\mathcal{G}$, i.e.,*

$$\mathcal{F}(x) \subseteq \mathcal{G}(x), \quad \forall x \in M.$$

*Furthermore, it satisfies that*

$$\Gamma(x) \subseteq \overline{\Gamma}(\mathcal{F}(x)).$$

**Proof.** Let $\mathcal{G}$ be a foliation of $M$ in the condition of the corollary. Then, we consider the family of manifolds given by the $\beta$-fibres $\Gamma(x)^x$. By left translations we generate a foliation of $\Gamma$ into submanifolds. By using Corollary 6.1.12 we have finished. $\qquad\square$

As a particular consequence we have that: $\overline{\Gamma}$ *is a transitive Lie subgroupoid of $\Gamma$ if and only if $M = \mathcal{F}(x)$ and $\overline{\Gamma} = \overline{\Gamma}(\mathcal{F}(x))$ for some $x \in M$.*

Let us give another consequence. Define the equivalence relation $\sim$ on $M$ given by

$$x \sim y \Leftrightarrow \exists \bar{g} \in \overline{\Gamma}, \quad \alpha(\bar{g}) = x, \ \beta(\bar{g}) = y.$$

Then, Example 6.1.8 provides another integrable distribution $A\overline{\Gamma}^{B}$ at $M$ called *transitive distribution of* $\overline{\Gamma}$. The associated foliation $\mathcal{G}$ of $M$ will be called *transitive foliation of* $\overline{\Gamma}$.

**Corollary 6.1.15.** *The base-characteristic foliation $\mathcal{F}$ based on the groupoid $\overline{\Gamma}$ is contained in the foliation $\mathcal{G}$ based on the equivalence relation $\sim$.*

**Proof.** For each $x \in M$, it satisfies that $\mathcal{F}(x) \times \mathcal{F}(x)$ defines a transitive Lie subgroupoid of $M \times M$ over $\mathcal{F}(x)$. So, using Corollary 6.1.14 we have done. $\qquad\square$

Thus, given a subgroupoid $\overline{\Gamma}$ of a Lie groupoid $\Gamma$ we have defined three canonical foliations, $\overline{\mathcal{F}}$, $\mathcal{F}$ and $\mathcal{G}$. Roughly speaking, $\mathcal{G}$ divides the base manifold into a maximal foliation such that each leaf is transitive. The main difference between the foliations $\mathcal{G}$ and $\mathcal{F}$ is that, with $\mathcal{F}$, we are not only requesting *"regularity"* on the base manifold $M$ but also on the groupoid $\overline{\Gamma}$. In particular, assume that $\overline{\Gamma}$ is a transitive subgroupoid of $\Gamma$. Then, $\mathcal{G}$ consists in one unique leaf equal to $M$. However, if $\overline{\Gamma}$ is not a Lie subgroupoid of $\Gamma$ the characteristic foliation $\overline{\mathcal{F}}$ is not given by the $\overline{\beta}$-fibres and, hence, the base-characteristic foliation $\mathcal{F}$ does not have (necessarily) one unique leaf equal to $M$.

We will now study another example. Let $\pi : A \rightarrow M$ be a vector bundle. Then, we may consider a structure of Lie groupoid $A \rightrightarrows M$ where $\alpha = \beta = \pi$ and the composition law is given by the sum of vectors of the fibres of $A$. So, a subgroupoid $E \rightrightarrows M$ of $A$ is simply a collection of vector subspaces $E_x$ of each fibre $A_x$ at the points $x \in M$.

Therefore, we may consider our characteristic and base-characteristic foliations $\overline{\mathcal{F}}(v_x)$ and $\mathcal{F}(x)$, respectively. However, in this case for all $x \in M$

$$\beta^{-1}(x) = \alpha^{-1}(x) = A_x.$$

Then, $\mathcal{F}(x) = \{x\}$. On the other hand,

$$\overline{\mathcal{F}}(0_x) = E_x, \quad \forall x \in M \tag{6.13}$$

if and only if $E$ is smooth (in the sense of that any vector in $E$ may be locally extended to a local section tangent to $E$). In fact, suppose that Eq. (6.13) is satisfied and fix $x \in M$. Let be a vector $v_x \in E_x$. Hence, by construction, we may assume that there exists an admissible vector field $\Theta$ such that its flow satisfies that $\varphi_{t_0}^\Theta (0_x) = v_x$ for some $t_0$. Thus, the (local) section $\sigma$ of $A$ given by

$$\sigma(y) = \varphi_{t_0}^\Theta (0_y),$$

is tangent to $E$ and $\sigma(x) = v_x$. The converse is analogous.

So, this model of groupoid does not provide any significant information about the collection $E$ of vector subspaces of $A$.

On the other hand, let us consider the frame groupoid $\Phi(A)$ on $A$. Then, we may construct the following subgroupoid:

$$\Phi_E(A) := \{ L_x^y \in \Phi(A)_x^y : L_x^y (E_x) = E_y \}.$$

Obviously, by arguments of dimensionality, $\Phi_E(A)$ is a Lie subgroupoid of $\Phi(A)$ only when it is transitive, i.e., all the vector spaces $E_x$ have the same dimension. Furthermore, we may prove that $\Phi_E(A)$ is a Lie subgroupoid of $\Phi(A)$ if and only if $E$ is a vector subbundle of $A$.

Then, by using Theorem 6.1.13, we may prove the following theorem.

**Theorem 6.1.16.** *Let* $\pi : A \to M$ *be a vector bundle. Consider a collection of vector subspaces* $E_x$ *of each fibre* $A_x$ *at the points* $x \in M$. *Then, there exists a maximal foliation of* $\mathcal{F}$ *of* $M$ *such that the union*

$$E_{\mathcal{F}(x)} = \bigsqcup_{y \in \mathcal{F}(x)} E_y,$$

*is a vector subbundle of* $A$.

So, roughly speaking, any collection of vector subspaces of a vector bundle may be separated into a foliation of vector subbundles in a maximal way.

## 6.2 Uniformity and Homogeneity

In this section, we will apply the results of the previous chapter to the case of simple materials. Thus, let $\mathcal{B}$ be a simple body with

$W : \Pi^1(\mathcal{B}, \mathcal{B}) \to V$ as the mechanical response (see Chapter 2). Then, we may define the so-called material groupoid $\Omega(\mathcal{B})$ which is a subgroupoid of the groupoid of 1-jets $\Pi^1(\mathcal{B}, \mathcal{B})$. So, it makes sense to apply here the development of Section 6.1. Let $\Theta$ be an admissible left-invariant vector field on $\Pi^1(\mathcal{B}, \mathcal{B})$, i.e., $\varphi_t^\Theta(\epsilon(X)) \in \Omega(\mathcal{B})$ for all $X \in \mathcal{B}$ and $t$ in the domain of the flow at $\epsilon(X)$. Then, for all $g \in \Pi^1(\mathcal{B}, \mathcal{B})$, we have that

$$TW(\Theta(g)) = \frac{\partial}{\partial t_{|0}}\left(W\left(\varphi_t^\Theta(g)\right)\right)$$

$$= \frac{\partial}{\partial t_{|0}}\left(W\left(g \cdot \varphi_t^\Theta(\epsilon(\alpha(g)))\right)\right)$$

$$= \frac{\partial}{\partial t_{|0}}\left(W(g)\right) = 0.$$

Therefore, analogously to the case of the material algebroid (see Eq. (5.12) in Chapter 5), we have that

$$TW(\Theta) = 0. \tag{6.14}$$

The converse is proved in the same way.

So, the characteristic distribution $A\Omega(\mathcal{B})^T$ of the material groupoid is generated by the (left-invariant) vector fields on $\Pi^1(\mathcal{B}, \mathcal{B})$ which are in the kernel of $TW$. This characteristic distribution will be called *material distribution*. The base-characteristic distribution $A\Omega(\mathcal{B})^\sharp$ will be called *body-material distribution* and the transitive distribution will be called *uniform-material distribution*. Let us recall that the left-invariant vector fields on $\Pi^1(\mathcal{B}, \mathcal{B})$ which satisfy Eq. (6.14) are called admissible vector fields and the family of these vector fields is denoted by $\mathcal{C}$.

Denote by $\overline{\mathcal{F}}(\epsilon(X))$, $\mathcal{F}(X)$ and $\mathcal{G}(X)$ the foliations associated to the material distribution, the body-material distribution and uniform-material foliation, respectively. For each $X \in \mathcal{B}$, we will denote the Lie groupoid $\Omega(\mathcal{B})(\mathcal{F}(X))$ by $\Omega(\mathcal{F}(X))$ (see Theorem 6.1.13). Denote the groupoid of all material isomorphisms at points in $\mathcal{G}(X)$ by $\Omega(\mathcal{G}(X))$. Recall that $\Omega(\mathcal{F}(X))$ is a subgroupoid of $\Omega(\mathcal{G}(X))$, i.e., $\Omega(\mathcal{F}(X)) \leq \Omega(\mathcal{G}(X))$.

### Graded uniformity

Notice that, strictly speaking, in continuum mechanics a *sub-body* of a body $\mathcal{B}$ is an open submanifold of $\mathcal{B}$ but, here, the foliation $\mathcal{F}$ gives us submanifolds of different dimensions. So, we will consider a more general definition so that, a *material submanifold (or generalized sub-body) of* $\mathcal{B}$ is just a submanifold of $\mathcal{B}$. A generalized sub-body $\mathcal{P}$ inherits certain material structure from $\mathcal{B}$. In fact, we will measure the material response of a material submanifold $\mathcal{P}$ by restricting $W$ to the 1-jets of local diffeomorphisms $\phi$ on $\mathcal{B}$ from $\mathcal{P}$ to $\mathcal{P}$. However, it is easy to observe that a material submanifold of a body is not exactly a body. See Jiménez *et al.* (2017) for a discussion on this subject.

Then, as a corollary of Theorem 6.1.3 and Corollary 6.1.14, we have the following result.

**Theorem 6.2.1.** *For all $X \in \mathcal{B}$, $\Omega\left(\mathcal{F}\left(X\right)\right)$ is a transitive Lie subgroupoid of $\Pi^{1}\left(\mathcal{B}, \mathcal{B}\right)$. Thus, any body $\mathcal{B}$ can be covered by a maximal foliation of smoothly uniform material submanifolds.*

Notice that, in this case "maximal" means that any other foliation $\mathcal{H}$ by smoothly uniform material submanifolds is thinner than $\mathcal{F}$, i.e.,

$$\mathcal{H}\left(X\right) \subseteq \mathcal{F}\left(X\right), \quad \forall X \in \mathcal{B}.$$

So, by using the material distributions we have been able to prove a very intuitive result: Let $\mathcal{B}$ be a general (smoothly uniform or not) simple material. Then, $\mathcal{B}$ may be decomposed into *"smoothly uniform parts"* and this decomposition is, in fact, a foliation of the material body.

We can ask now the same question for (not generally smooth) uniformity. To solve this problem, we will take advantage of the uniform-material distribution $A\overline{\Omega}^{B}\left(\mathcal{B}\right)$. Therefore, by using Corollary 6.1.14, we obtain the similar result which we were looking for.

**Theorem 6.2.2.** *Any simple body $\mathcal{B}$ can be covered by a maximal foliation $\mathcal{G}$ of uniform material submanifolds.*

In this case, the maximality has the same meaning. Note that Corollary 6.1.15 provides a very intuitive results in this context: *The uniform leaves are generally bigger than smoothly uniform leaves.*

**Remark 6.2.3.** Imagine that there is, at least, a 1-jet $g \in \Omega^X(\mathcal{B})$ for some $X \in \mathcal{B}$ such that

$$g \notin \overline{\mathcal{F}}(\epsilon(X)).$$

Then, we are not including $g$ inside any of the transitive Lie subgroupoids $\Omega(\mathcal{F}(X))$. Thus, these material isomorphisms are being discarded.

Nevertheless

$$\overline{\mathcal{F}}(g) = g \cdot \overline{\mathcal{F}}(\epsilon(\alpha(g))), \tag{6.15}$$

and, indeed, $\overline{\mathcal{F}}(\epsilon(\alpha(g)))$ is contained in $\Omega(\mathcal{F}(\alpha(g)))$, i.e., using Eq. (6.15), we can reconstruct $\overline{\mathcal{F}}(g)$.

Finally, using the body-material distribution, we will be able to define a more general notion of (smooth) uniformity. The generalization of smooth uniformity was introduced in Epstein *et al.* (2019). We will end up using the foliation by (smoothly) uniform sub-bodies to interpret it over the material groupoid.

**Definition 6.2.4.** Let $\mathcal{B}$ be a body and $X \in \mathcal{B}$ be a body point. Then, $\mathcal{B}$ is said to be *smoothly uniform of grade $p$ at $X$* if $A\Omega(\mathcal{B})_X^\sharp$ has dimension $p$. $\mathcal{B}$ is *smoothly uniform of grade $p$* if it is smoothly uniform of grade $p$ at all the points.

On the other hand, $\mathcal{B}$ is said to be *uniform of grade $p$ at $X$* if $A\Omega(\mathcal{B})_X^B$ has dimension $p$. $\mathcal{B}$ is *uniform of grade $p$* if it is uniform of grade $p$ at all the points.

Note that, smooth uniformity (respectively, uniformity) is a particular case of graded smooth uniformity (respectively, graded uniformity). In fact, $\mathcal{B}$ is (smoothly) uniform if and only if $\mathcal{B}$ is (smoothly) uniform of grade 3. So, $\mathcal{B}$ is (smoothly) uniform of grade 3 if and only if $(A\Omega(\mathcal{B})_X^\sharp) A\Omega(\mathcal{B})_X^B$ has dimension 3 for all $X \in \mathcal{B}$, i.e., there exists just one leaf of the (body-material) uniform-material foliation equal to $\mathcal{B}$. Equivalently, the material groupoid $\Omega(\mathcal{B})$ is a (Lie) transitive subgroupoid of $\Pi^1(\mathcal{B}, \mathcal{B})$ whose $\overline{\beta}$-fibres integrate the material distribution.

**Corollary 6.2.5.** *Let $\mathcal{B}$ be a body and let $X \in \mathcal{B}$ be a body point. Then $\mathcal{B}$ is smoothly uniform of grade $p$ at $X$ if and only if the smoothly uniform leaf $\mathcal{F}(X)$ at $X$ has dimension $p$.*

*On the other hand, $\mathcal{B}$ is uniform of grade $p$ at $X$ if and only if the uniform leaf $\mathcal{G}(X)$ at $X$ has dimension $p$.*

**Corollary 6.2.6.** *Let $\mathcal{B}$ be a body. Then $\mathcal{B}$ is (smoothly) uniform of grade $p$ if and only if the (body-material) uniform-material foliation is regular of rank $p$.*

It is important to highlight again that the body-material foliation and the uniform-material foliation have certain condition of maximality (see Theorem 6.2.1). In fact, suppose that there exists another foliation $\mathcal{H}$ of $\mathcal{B}$ by (smoothly) uniform material submanifolds. Then, for all $X \in \mathcal{B}$ we have that

$$\mathcal{H}(X) \subseteq (\mathcal{F}(X))\mathcal{G}(X), \quad \forall X \in \mathcal{B}.$$

So, we have the following results.

**Corollary 6.2.7.** *Let $\mathcal{B}$ be a body and let $X \in \mathcal{B}$. Then $\mathcal{B}$ is (smoothly) uniform of grade greater than or equal to $p$ at $X$ if and only if there exists a foliation $\mathcal{H}$ of $\mathcal{B}$ by (smoothly) uniform submanifolds such that the leaf $\mathcal{H}(X)$ at $X$ has dimension greater than or equal to $p$.*

**Corollary 6.2.8.** *Let $\mathcal{B}$ be a body. Then $\mathcal{B}$ is (smoothly) uniform of grade $p$ if and only if the body can be foliated by (smoothly) uniform material submanifolds of dimension $p$.*

### Homogeneity

This section is devoted to deal with the definition of homogeneity. As we already know, a body is (locally) homogeneous if it admits a (local) configuration $\phi$ which induces a left (local) smooth field of material isomorphisms $\mathcal{P}$ given by

$$\mathcal{P}(Y, X) = j_{Y,X}^1 \left( \phi^{-1} \circ \tau_{\phi(X) - \phi(Y)} \circ \phi \right), \tag{6.16}$$

for all body point $Y$ in the domain of $\phi$ and a fixed $X \in \mathcal{B}$ (see Definition 2.0.8). Roughly speaking, a body is said to be homogeneous if we can choose a section of the material groupoid which is constant on the body. As we have said before, local homogeneity is clearly more restrictive than smooth uniformity. In fact, in this case, the smooth fields of material isomorphisms (see Definition 4.0.46) are induced by particular (local) configurations.

However, in a purely intuitive picture, homogeneity can be interpreted as the absence of defects. So, it makes sense to develop a concept of some kind of homogeneity for non-uniform materials which measures the absence of defects and generalizes the known one. In the literature we can already find some partial answer of this question (Cámpos *et al.*, 2008; Epstein and de León, 2000 for FGM's and Epstein, 2017; Epstein *et al.*, 2019 for laminated and bundle materials).

Recall that the material distributions are characterized by the commutativity of the following diagram:

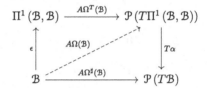

As we have proved in the previous chapter, the body-material foliation $\mathcal{F}$ divides the body into smoothly uniform components.

Let us now provide the intuition behind the definition of homogeneity of a non-uniform body. A non-uniform body will be (*locally*) *homogeneous* when each smoothly uniform material submanifold $\mathcal{F}(X)$ is (locally) homogeneous and all the uniform material submanifolds can be straightened at the same time.

Thus, we need to clarify what we understand by homogeneity of submanifolds of $\mathcal{B}$.

**Definition 6.2.9.** Let $\mathcal{B}$ be a simple body and $\mathcal{N}$ be a submanifold of $\mathcal{B}$. Then $\mathcal{N}$ is said to be *homogeneous* if and only if for all point $X \in \mathcal{N}$ there exists a local configuration $\psi$ of $\mathcal{B}$ on an open subset $U \subseteq \mathcal{B}$, with $\mathcal{N} \subseteq U$, which satisfies that

$$j^1_{Y,Z}\left(\psi^{-1} \circ \tau_{\psi(Z)-\psi(Y)} \circ \psi\right),$$

is a material isomorphism for all $Y, Z \in \mathcal{N}$. We will say that $\mathcal{N}$ is *locally homogeneous* if there exists a covering of $\mathcal{N}$ by open subsets $U_a$ of $\mathcal{B}$ such that $U_a \cap \mathcal{N}$ are *homogeneous* submanifolds of $\mathcal{B}$. $\mathcal{N}$ is said to be (*locally*) *inhomogeneous* if it is not (*locally*) *homogeneous*.

Notice that, the definitions of homogeneity and local homogeneity for smoothly uniform materials (Definition 2.0.8) are generalized by this one whether $\mathcal{N} = \mathcal{B}$ or $\mathcal{N}$ is just an open subset of $\mathcal{B}$.

Now, taking into account that $\mathcal{F} = \{\mathcal{F}(X)\}_{X \in \mathcal{B}}$ is a foliation, there is a kind of compatible atlas which is called a foliated atlas (see Appendix A). In fact, $\{((x_a^i), U_a)\}_a$ is a foliated atlas of $\mathcal{B}$ associated to $\mathcal{F}$ whenever for each $X \in U_a \subseteq \mathcal{B}$ we have that $U_a := \{-\epsilon < x_a^1 < \epsilon, \ldots, -\epsilon < x_a^3 < \epsilon\}$ for some $\epsilon > 0$, such that the $k$-dimensional disk $\{x_a^{k+1} = \cdots = x_a^3 = 0\}$ coincides with the path-connected component of the intersection of $\mathcal{F}(X)$ with $U_a$ which contains $X$, and each $k$-dimensional disk $\{x_a^{k+1} = c_{k+1}, \ldots, x_a^3 = c_3\}$, where $c_{k+1}, \ldots, c_3$ are constants, is wholly contained in some leaf of $\mathcal{F}$. Intuitively, this atlas straightens (locally) the partition $\mathcal{F}$ of $\mathcal{B}$.

The existence of these kinds of atlases and the maximality condition over the smoothly uniform material submanifolds $\mathcal{F}(X)$ induces us to give the following definition.

**Definition 6.2.10.** Let $\mathcal{B}$ be a simple body. Then $\mathcal{B}$ is said to be *locally homogeneous* if and only if for all point $X \in \mathcal{B}$ there exists a local configuration $\psi$ of $\mathcal{B}$, with $X \in U$, which is a foliated chart and it satisfies that

$$j_{Y,Z}^1 \left( \psi^{-1} \circ \tau_{\psi(Z)-\psi(Y)} \circ \psi \right),$$

is a material isomorphism for all $Z \in U \cap \mathcal{F}(Y)$. We will say that $\mathcal{B}$ is homogeneous if $U = \mathcal{B}$. The body $\mathcal{B}$ is said to be (*locally*) *inhomogeneous* if it is not (locally) homogeneous.

It is remarkable that, as we have said above, all the uniform leaves $\mathcal{F}(X)$ of a homogeneous body are homogeneous. Therefore, the definition of homogeneity for a smoothly uniform body coincides with Definition 2.0.8. Notice also that, the condition that all the leaves $\mathcal{F}(X)$ are homogeneous is not enough in order to have the homogeneity of the body $\mathcal{B}$ because there is also a condition of compatibility with the foliation structure of $\mathcal{F}$.

Let us recall a result presented in Chapter 2 given in Elżanowski *et al.* (1990) (see also Elżanowski and Prishepionok, 1992 or Wang, 1967) which characterizes the homogeneity by using $G$-structures.

Fix $\bar{g}_0$ be a frame at $Z_0 \in \mathcal{B}$. Then, assuming that $\mathcal{B}$ is smoothly uniform, the set

$$\Omega\left(\mathcal{B}\right)_{Z_0} \cdot \bar{g}_0 := \{\bar{g} \cdot \bar{g}_0 : \bar{g} \in \Omega\left(\mathcal{B}\right)_{Z_0}\},$$

where $\cdot$ defines the composition of 1-jets, is a $G_0$-structure over $\mathcal{B}$ where

$$G_0 := \overline{Z}_0^{-1} \cdot G\left(Z_0\right) \cdot \overline{Z}_0.$$

**Proposition 6.2.11.** *Let $\bar{g}_0 \in F\mathcal{B}$ be a frame. If $\mathcal{B}$ is homogeneous, then the $G_0$-structure given by $\Omega\left(\mathcal{B}\right) \cdot \bar{g}_0$ is integrable. Conversely, $\Omega\left(\mathcal{B}\right) \cdot \bar{g}_0$ is integrable implies that $\mathcal{B}$ is locally homogeneous.*

Thus, the next step will be to give a similar result for this generalized homogeneity. Because of the lack of uniformity we have to use groupoids instead of $G$-structures.

Let $\mathbb{S} := \{\mathbb{S}\left(x\right) : x \in \mathbb{R}^n\}$ be a canonical foliation of $\mathbb{R}^n$ (see Example A.0.6), i.e., for all $x = \left(x^1, \ldots, x^n\right) \in \mathbb{R}^n$ the leaf $\mathbb{S}\left(x\right)$ at $x$ is given by

$$\mathbb{S}\left(x\right) := \{\left(y^1, \ldots, y^p, x^{p+1}, \ldots, x^n\right) : y^i \in \mathbb{R}, \ i = 1, \ldots, p\},$$

for some $1 \leq p \leq n$.

Recall that for any foliation $\mathcal{G}$ on a manifold $Q$ there exists a map

$$\dim_{\mathcal{G}} : Q \to \{0, \ldots, \dim\left(Q\right)\},$$

such that for all $x \in Q$

$$\dim_{\mathcal{G}}\left(x\right) = \dim\left(\mathcal{G}\left(x\right)\right).$$

It is important to remark that in the case of $\mathbb{S}$ the dimension $\dim_{\mathbb{S}}$ characterizes the foliation $\mathbb{S}$. Thus, with abuse of notation, we could say that the map $\dim_{\mathbb{S}}$ is the foliation.

Let $\mathbb{S}$ be a canonical foliation of $\mathbb{R}^n$ with dimension $p = \dim_{\mathbb{S}}$. Then, as a generalization of the frame bundle of $\mathbb{R}^n$, we define the *p-graded frame groupoid* as the following subgroupoid of $\Pi^1\left(\mathbb{R}^n, \mathbb{R}^n\right)$,

$$\Pi_p^1\left(\mathbb{R}^n, \mathbb{R}^n\right) = \{j_{x,y}^1 \psi \in \Pi^1\left(\mathbb{R}^n, \mathbb{R}^n\right) : y \in \mathbb{S}\left(x\right)\}.$$

Notice that the restriction of $\Pi_p^1\left(\mathbb{R}^n, \mathbb{R}^n\right)$ to any leaf $\mathbb{S}\left(x\right)$ is a transitive Lie subgroupoid of $\Pi^1\left(\mathbb{R}^n, \mathbb{R}^n\right)$ with all the isotropy groups isomorphic to $Gl(n, \mathbb{R})$. However, the groupoid $\Pi_p^1(\mathbb{R}^n, \mathbb{R}^n)$ is not necessarily a Lie subgroupoid of $\Pi^1(\mathbb{R}^n, \mathbb{R}^n)$. In fact, $\Pi_p^1(\mathbb{R}^n, \mathbb{R}^n)$ *is a Lie subgroupoid of* $\Pi^1(\mathbb{R}^n, \mathbb{R}^n)$ *if and only if* $\mathbb{S}$ *is regular foliation.*

A *standard flat G-reduction of grade* $p$ is a subgroupoid $\Pi_{G,p}^1(\mathbb{R}^n, \mathbb{R}^n)$ of $\Pi_p^1(\mathbb{R}^n, \mathbb{R}^n)$ such that the restrictions $\Pi_{G,p}^1(\mathbb{S}(x), \mathbb{S}(x))$ to the leaves $\mathbb{S}(x)$ are transitive Lie subgroupoids of $\Pi^1(\mathbb{R}^n, \mathbb{R}^n)$ on the leaf $\mathbb{S}(x)$. It is remarkable that in this case all the isotropy groups of $\Pi_{G,p}^1(\mathbb{S}(x), \mathbb{S}(x))$ are conjugated.

Clearly, all the structures introduced in this section can be restricted to any open subset of $\mathbb{R}^n$.

Let $\psi : U \to \overline{U}$ be a (local) configuration on $U \subseteq \mathcal{B}$. Then, $\psi$ induces a Lie-groupoid isomorphism,

$$\Pi\psi : \Pi^1(U, U) \to \Pi^1(\overline{U}, \overline{U})$$

$$j_{X,Y}^1 \phi \mapsto j_{\psi(X), \psi(Y)}^1 \left( \psi \circ \phi \circ \psi^{-1} \right)$$

**Proposition 6.2.12.** *Let $\mathcal{B}$ be a simple body. If $\mathcal{B}$ is homogeneous, the material groupoid is isomorphic (via a global configuration) to a standard flat G-reduction. Conversely, if the material groupoid is isomorphic (via a local configuration) to a standard flat G-reduction, then $\mathcal{B}$ is locally homogeneous.*

Notice that, in the context of principal bundles, a $G$-structure is integrable if and only if there exists a local configuration which induces an isomorphism from the $G$-structure to a standard flat $G$-structure. Observe also that this results in a natural generalization of Proposition 5.1.5. In fact, implicitly, we are generalizing the notion of integrability definition (Definition 5.1.1).

Finally, we will use the material distribution to give another characterization of homogeneity.

Let $\mathcal{B}$ be a homogeneous body with $\psi = (x^i)$ as a (local) homogeneous configuration. Then, by using the fact that $\psi$ is a foliated chart, we have that the partial derivatives are tangent to $A\Omega(\mathcal{B})^\sharp$, i.e., for each $X \in U$

$$\frac{\partial}{\partial x^l}_{|X} \in A\Omega(\mathcal{B})_X^\sharp,$$

for all $1 \leq l \leq \dim (\mathcal{F}(X)) = K$. Thus, there are local functions $\Lambda_{i,j}^j$ such that for each $l \leq K$ the (local) left-invariant vector field on $\Pi^1(\mathcal{B},\mathcal{B})$ given by

$$\frac{\partial}{\partial x^l} + \Lambda_{i,l}^j \frac{\partial}{\partial y_i^j},$$

is tangent to $A\Omega(\mathcal{B})^T$, where $(x^i, y^j, y_i^j)$ are the induced coordinates of $(x^i)$ in $\Pi^1(\mathcal{B},\mathcal{B})$. Equivalently, the local functions $\Lambda_{i,l}^j$ satisfy that

$$\frac{\partial W}{\partial x^l} + \Lambda_{i,l}^j \frac{\partial W}{\partial y_i^j} = 0,$$

for all $1 \leq l \leq K$. Next, since for each two points $X, Y \in U$ the 1-jet given by $j_{X,Y}^1 \left( \psi^{-1} \circ \tau_{\psi(Y)-\psi(X)} \circ \psi \right)$ is a material isomorphism, we can choose $\Lambda_{i,l}^j = 0$.

**Proposition 6.2.13.** *Let $\mathcal{B}$ be a simple body. Then $\mathcal{B}$ is homogeneous if and only if for each $X \in \mathcal{B}$ there exists a local chart $(x^i)$ on $\mathcal{B}$ at $X$ such that,*

$$\frac{\partial W}{\partial x^l} = 0, \qquad\qquad (6.17)$$

*for all $l \leq \dim (\mathcal{F}(X))$.*

Notice that Eq. (6.17) implies that the partial derivatives of the coordinates $(x^i)$ up to $\dim (\mathcal{F}(X))$ are tangent to the material distribution and, therefore, the coordinates are foliated. So, Eq. (6.17) gives us an apparently more straightforward way to express this "*generalized*" homogeneity.

## 6.3   Examples

We will devote this chapter to apply the notions of graded uniformity (Definition 6.2.4) and homogeneity (Definition 6.2.10) for non-uniform bodies. In particular, we will present two (family of) examples: a homogeneous non-uniform body and a (generally) inhomogeneous non-uniform body.

We will see that, in some of them, the material groupoid is not a Lie groupoid (these kinds of examples justify the study of groupoids without structure of Lie groupoids). We shall also give the decomposition of the material by (smoothly) uniform material submanifolds provided by the characteristic distribution. In these examples, we will also show that the leaves $\overline{\mathcal{F}}\left(\epsilon\left(X\right)\right)$ are contained in the $\overline{\beta}$-fibres of $\Omega\left(\mathcal{B}\right)$ but they do not coincide in general.

### Example 1

Let $\mathcal{B}$ be a simple material body for which there exists a reference configuration $\psi_0$ from $\mathcal{B}$ to the three-dimensional open ball $B_r\left(0\right)$ of centre 0 and radius $r > 1$ in $\mathbb{R}^3$ which induces the following mechanical response:

$$W : \Pi^1\left(B_r\left(0\right), B_r\left(0\right)\right) \to \mathfrak{gl}\left(3, \mathbb{R}\right)$$
$$j^1_{X,Y}\phi \mapsto f\left(\parallel X \parallel^2\right)\left(F \cdot F^T - I\right),$$

such that

$$f\left(t\right) = \begin{cases} 1 & \text{if } t \leq 1, \\ 1 + e^{-\frac{1}{t-1}} & \text{if } t > 1, \end{cases}$$

where $\mathfrak{gl}\left(3, \mathbb{R}\right)$ is the algebra of matrices, $F$ is the Jacobian matrix of $\phi$ at $X$ with respect to the canonical basis of $\mathbb{R}^3$ and $I$ is the identity matrix. Here, the (global) canonical coordinates of $\mathbb{R}^3$ are denoted by $\left(X^I\right)$ and $X = \left(X^1, X^2, X^3\right)$ with respect to these coordinates. In these coordinates, we allow the summation convention to be in force regardless of the placement of the indices. We also identify the coordinate system in the spatial configuration with that of the reference configuration.

Notice that $f$ is constant up to 1 and strictly increasing thereafter. For this reason, one can immediately conclude that $B_r\left(0\right)$ is not uniform. In fact, there are no material isomorphisms joining any two points $X$ and $Y$ such that

$$f\left(\parallel X \parallel^2\right) \neq f\left(\parallel Y \parallel^2\right).$$

So, let us study the derivatives of $W$ in order to find the grades of uniformity of the points of the body $B_r(0)$. We obtain

$$\frac{\partial W^{ij}}{\partial F_M^k} = f\left(\|X\|^2\right)[\delta_k^i F_M^j + \delta_k^j F_M^i],$$

$$\frac{\partial W^{ij}}{\partial X^I} = 2X^I \frac{\partial f}{\partial t}(F_K^i F_K^j - \delta^{ij}).$$

We are looking for left-invariant (local) vector fields $\Theta$ on $\Pi^1(B_r(0), B_r(0))$ satisfying

$$\Theta\left(W^{ij}\right) = 0. \tag{6.18}$$

Let $(X^I, Y^I, F_j^i)$ be the induced coordinates of $(X^I)$ on $\Pi^1(B_r(0), B_r(0))$. A left-invariant vector field $\Theta$ can be expressed as follows:

$$\Theta(X^I, Y^J, F_I^j) = ((X^I, Y^J, F_I^j), \delta X^I, 0, F_L^j \delta P_I^L).$$

Hence, $\Theta$ satisfies Eq. (6.18) if and only if

$$\Theta\left(W^{ij}\right) = f\left(\| X \|^2\right)(F_L^i F_M^j + F_L^j F_M^i)\delta P_M^L$$

$$+ 2X^I \delta X^I \frac{\partial f}{\partial t}(F_K^i F_K^j - \delta^{ij}) = 0.$$

Let us focus first on the open given by the restriction $\|X\|^2 < 1$. Then, the above equation turns as follows:

$$(F_L^i F_M^j + F_L^j F_M^i)\delta P_M^L = 0, \quad \forall i, j = 1, 2, 3 \tag{6.19}$$

for every Jacobian matrix $F = (F_L^j)$ of a local diffeomorphism $\phi$ on $B_r(0)$. Since the bracketed expression is symmetric in $L$ and $M$ for every $i$ and $j$, it follows that $\delta P$ is a skew-symmetric matrix. We remark that this condition does not impose any restriction on the components $\delta X^I$ of the admissible vector fields on the base vectors $\partial/\partial X^I$. In other words, any family of local functions $\{\delta X^I, \delta P_M^L\}$ on the open restriction $\{\|X\|^2 < 1\}$ of the body $B_r(0)$, such that $\delta P = (\delta P_M^L)$ is a skew-symmetric matrix, generates a vector field

$$\Theta(X^I, Y^J, F_I^j) = ((X^I, Y^J, F_I^j), \delta X^I, 0, F_L^j \delta P_I^L),$$

which satisfies Eq. (6.18). It follows that the body characteristic distribution of the sub-body $B_1(0)$ is a regular distribution of dimension 3. Therefore, this sub-body is smoothly uniform, as one would expect from the constancy of the function $f$. Note also that the part lost when projecting the characteristic distribution onto the body, namely the skew-symmetric matrices $\delta P$, consists precisely of the Lie algebra of the orthogonal group. This is nothing but the manifestation of the fact that our sub-body is isotropic.

Next we will study the open subset of $B_r(0)$ such that $\|X\|^2 > 1$. For this case, Eq. (6.18) is satisfied if and only if

$$f\left(\|X\|^2\right)(F_L^i F_M^j + F_L^j F_M^i)\delta P_M^L + 2X^I \delta X^I \frac{\partial f}{\partial t} F_K^i F_K^j$$

$$= 2X^I \delta X^I \frac{\partial f}{\partial t}\delta^{ij}.$$

The function on the left-hand side of this equation is homogeneous of degree 2 with respect to the matrix coordinate $F$, but the function on the right-hand side does not depend on $F$. Consequently, the above equation can be identically satisfied if and only if

$$\delta X^I \frac{\partial f}{\partial t} = 0, \quad \forall I. \tag{6.20}$$

Notice that, the map $f$ is strictly monotonic (and, hence, a submersion) at the open subset given by the condition $\|X\|^2 > 1$. Then, for any point $X$ in this open subset we have that

$$T_X f^{-1}\left(f\left(\|X\|^2\right)\right) = \text{Ker}\left(T_X f\right),$$

i.e., the tangent space of the level set $f^{-1}\left(f\left(\|X\|^2\right)\right)$, which is the sphere $\|Y\|^2 = \|X\|^2$, consists of vectors $V = \left(V^1, V^2, V^3\right)$ such that

$$V^1 \frac{\partial f}{\partial t_{|\|X\|^2}} = 0.$$

In this way, a vector field $\Theta$ satisfies Eq. (6.18) if and only if $\delta P$ is skew-symmetric and the projection $T\alpha \circ \Theta \circ \epsilon$ is tangent to the vertical spheres $\|Y\|^2 = C$. Therefore, for each point $X = \left(X^1, X^2, X^3\right)$ with $\|X\|^2 > 1$, the smoothly uniform leaf is given by the sphere

$$\|Y\|^2 = \|X\|^2.$$

Fig. 6.1.   Material foliation of $B_r(0)$.

As a consequence, the smoothly uniform leaf at the points satisfying $\|X\|^2 = 1$ is, again, the sphere $\|Y\|^2 = 1$. It is an exercise to prove that the uniform leaves are exactly the same.

We conclude that the body is (smoothly) uniform of grade 3 for all points $X = \left(X^1, X^2, X^3\right) \in B_r(0)$ such that $\|X\|^2 < 1$, and it is (smoothly) uniform of grade 2 otherwise. It should be remarked that the sphere $\|X\|^2 = 1$ is (smoothly) uniform of grade 2, even though its points are materially isomorphic to those in the subset with $\|X\|^2 < 1$ (see Fig. 6.1).

Finally, the material body $B_r(0)$ is locally homogeneous. In fact, let us consider the spherical coordinates $(r, \theta, \varphi)$ of $B_r(0)$. Then,

$$\frac{\partial W^{ij}}{\partial \theta} = \frac{\partial W^{ij}}{\partial \varphi} = 0,$$

i.e., by using Proposition 6.2.13, $B_r(0)$ is homogeneous and the coordinates $(r, \theta, \varphi)$ are homogeneous coordinates.

## *Example 2*

We will consider a perturbation of the model introduced by Coleman Coleman (1965) and Wang (1965) called *simple liquid crystal* introduced in Chapter 4. These kinds of materials will be called *laminated simple liquid crystals*.

In this case, we will consider a simple body $\mathcal{B}$ together with a reference configuration $\psi_1$ from $\mathcal{B}$ the open ball $\mathcal{B}_1 = B_r(0)$ in $\mathbb{R}^3$ of radius $r$ and centre $0 \in \mathbb{R}^3$. Furthermore, $\psi_1$ induces on $\mathcal{B}_1$ a mechanical response $\mathcal{W}$ determined by the following objects:

**(i)** A fixed vector field $e$ on $\mathcal{B}_1$ such that $e(X) \neq 0$ for all $X \in \mathcal{B}_1$.

**(ii)** Two differentiable maps $r, J : \Pi^1(\mathcal{B}_1, \mathcal{B}_1) \to \mathbb{R}$ in the following way:

- $r(j^1_{X,Y}\phi) = g(Y)(T_X\phi(e(X)), T_X\phi(e(X))) + \|X\|^2$,
- $J(j^1_{X,Y}\phi) = \det(F)$,

where $F$ is the Jacobian matrix of $\phi$ with respect to the canonical basis of $\mathbb{R}^3$ at $X$, $g$ is a Riemannian metric on $\mathcal{B}_1$ and $\|\cdot\|$ the Euclidean norm of $\mathbb{R}^3$.

**(iii)** A differentiable map $\widehat{W} : \mathbb{R}^2 \to V$, with $V$ a finite-dimensional $\mathbb{R}$-vector space.

Thus, these three objects induce a structure of simple body by considering the mechanical response $\mathcal{W} : \Pi^1(\mathcal{B}_1, \mathcal{B}_1) \to V$ as the composition

$$\mathcal{W} = \widehat{W} \circ (r, J).$$

Let us now fix the canonical (global) coordinates $(X^I)$ of $\mathbb{R}^3$. Then, these coordinates induce a (canonic) isomorphism $T\mathcal{B}_1 \cong \mathcal{B}_1 \times \mathbb{R}^3$. By using this isomorphism, any vector $V_X \in T_X\mathcal{B}_1$ can be equivalently expressed as $(X, V^i)$ in $\mathcal{B}_1 \times \mathbb{R}^3$. For the same reason, $r$ can be written as follows:

$$r(j^1_{X,Y}\phi) = g(Y)(F^j_L e^L(X), F^j_L e^L(X)) + \|X\|^2,$$

where $e(X) = (X, e^I(X)) \in \mathcal{B}_1 \times \mathbb{R}^3$. Both expressions will be used with the same notation as long as there is no confusion.

Now, we want to study the conditions characterizing the material distribution $A\Omega^T(\mathcal{B}_1)$. In particular, we should study the admissible left-invariant vector fields $\Theta$ on $\Pi^1(\mathcal{B}_1, \mathcal{B}_1)$, i.e.,

$$\Theta(\mathcal{W}) = 0. \tag{6.21}$$

Notice that, for each $U = (U^j_i) \in gl(3, \mathbb{R})$ and $v = (v^i) \in \mathbb{R}^3$, we have that,

**(i)** $\dfrac{\partial r}{\partial X^i_{|j^1_{X,Y}\phi}}(v) = 2g(Y)\left(F^j_L e^L(X), F^j_L \dfrac{\partial e^L}{\partial X^M_{|X}} v^M\right) + 2v^L X^L$,

**(ii)** $\dfrac{\partial r}{\partial F^i_{|j^1_{X,Y}\phi}}(U) = 2g(Y)(F^j_L e^L(X), U^j_L e^L(X))$,

**(iii)** $\dfrac{\partial J}{\partial F_{|j^1_{X,Y}\phi}}(U) = \det{(F)}\operatorname{Tr}\left(F^{-1}\cdot U\right).$

We are denoting the coordinate $X^K(X)$ by $X^K$.

Let $\left(X^I, Y^J, Y^J_I\right)$ be the induced local coordinates of the canonical coordinates $\left(X^I\right)$ of $\mathbb{R}^3$ in $\Pi^1(\mathcal{B}_1, \mathcal{B}_1)$. Then, $\Theta$ can be expressed as follows:

$$\Theta(X^I, Y^J, F^j_I) = ((X^I, Y^J, F^j_I), \delta X^I, 0, F^j_L \delta P^L_I).$$

Hence, $\Theta$ is an admissible vector field if and only if

$$0 = 2\frac{\partial \widehat{W}}{\partial r_{|j^1_{X,Y}\phi}} g(Y)\left(F^j_L e^L(X), F^j_L\frac{\partial e^L}{\partial X^M_{|X}}\delta X^M(X)\right)$$

$$+ 2\,\frac{\partial \widehat{W}}{\partial r_{|j^1_{X,Y}\phi}}\delta X^L(X)\,X^L$$

$$+ 2\,\frac{\partial \widehat{W}}{\partial r_{|j^1_{X,Y}\phi}} g(Y)\,(F^j_L e^L(X), F^j_L\delta P^L_M(X)\,e^M(X))$$

$$+ \det{(F)}\,\frac{\partial \widehat{W}}{\partial J_{|j^1_{X,Y}\phi}}\operatorname{Tr}(\delta P^j_I(X)),$$

for all $j^1_{X,Y}\phi \in \Pi^1(\mathcal{B}_1, \mathcal{B}_1)$. So, a sufficient but not necessary condition would be that, for each 1-jet of local diffeomorphisms $j^1_{X,Y}\phi$ on $\mathcal{B}_1$, it satisfies

**(1)** $\operatorname{Tr}(\delta P^j_I(X)) = 0;$

**(2)** denoting $\mathcal{L}^L_X = \dfrac{\partial e^L}{\partial X^M_{|X}}\delta X^M(X) + \delta P^L_M(X)\,e^M(X)$, then

$$g(Y)\,(F^j_L e^L(X), F^j_L\,(\mathcal{L}^L_X)) = -\delta X^L(X)\,X^L.$$

In order to turns this conditions into necessary conditions we will assume that $\widehat{W}$ is an immersion and, hence, **(1)** and **(2)** are equivalent to the above equation.

In this way, $\mathcal{B}_1$ is smoothly uniform if and only if for each vector $V_X$ at $X$ there exists a family of local functions $\{\delta X^I, \delta P^i_j\}$ at $X$ satisfying (1), (2) and

$$\delta X^I (X) = V^I, \quad \forall i,$$

where $V_X = (X, V^I) \in \mathcal{B}_1 \times \mathbb{R}^3$.

Let us focus on the second condition: Suppose that $\langle V^I, X \rangle \neq 0$. Then, fixing the spatial point $X \in \mathcal{B}_1$ the map depending on the matrix coordinates $F^j_i$,

$$g(Y) \left( F^j_L e^L (X), F^j_L \left( \frac{\partial e^L}{\partial X^M_{|X}} V^M + \delta P^l_M (X) e^M (X) \right) \right) \qquad (6.22)$$

is equal to $-\delta X^L (X) X^L$ which does not depend on the matrix coordinates $F^j_i$ and it is not zero. However, the map (6.22) depends bilinearly on $F^j_I$. So, Eq. (6.22) cannot be constant (respect to $F^j_i$) and different from zero at the same time. Therefore, we could conclude that $\mathcal{B}_1$ is not smoothly uniform.

This fact opens the possibility of studying the graduated (smooth) uniformity of these materials. Notice that, as we have proved, any admissible vector field $\Theta$ satisfies that

$$\delta X^L (X) X^L = 0, \qquad (6.23)$$

where $\Theta(X^I, Y^J, F^j_I) = ((X^I, Y^J, F^j_I), \delta X, 0, F^j_L \delta P^L_I)$ respect to the coordinates $(X^I, Y^J, F^j_I)$ on $\Pi^1(\mathcal{B}_1, \mathcal{B}_1)$.

Let $X \in \mathcal{B}_1$ be a point of the body different to 0. Then, the map given by $\| \cdot \|^2$ restricted to $\mathcal{B}_1$ has full rank at $X$. In fact, the level set of $\| \cdot \|^2$ at $\|X\|^2$ is given by the sphere $S(\|X\|)$ of radius $\|X\|$ and centre 0 and it satisfies that

$$T_X S(\|X\|) = \text{Ker} (T_X \| \cdot \|^2).$$

So, the tangent space of the sphere $S(\|X\|)$ at $X$ consists of the vectors $V_X = (X, V^I) \in \mathcal{B}_1 \times \mathbb{R}^3$ satisfying

$$V^L X^L = 0. \qquad (6.24)$$

Then, any vector $V_X$ satisfying Eq. (6.24) can be expanded by a (local) vector field $\theta^S$ on $S(\|X\|)$ such that

$$\theta^S (X) = V_X.$$

It is an easy exercise to prove that $\theta^S$ can be extended to a vector field $\theta$ on an open neighborhood $U$ of $\mathcal{B}_1$ which is tangent to all the spheres intersecting $U$. Then, expressing $\theta$ in the canonical coordinates $(X^I)$ as follows:

$$\theta\left(X^I\right) = \left(X^I, \delta X^I\right),$$

the functions $\delta X^i$ satisfy Eq. (6.23). Therefore, by using the non-degeneracy of the Riemannian metric $g$, it is enough to realize that there exist infinite families of local maps $\delta P_I^j$ at $X$ from the body to $\mathbb{R}$ satisfying that

**(1)″** $\delta P_I^i = 0$;
**(2)″** $\frac{\partial e^J}{\partial X^L} \delta X^L = -\delta P_L^j e^L,\ \forall j$,

Therefore, the local vector fields given by,

$$\Theta(X^I, Y^J, F_I^j) = ((X^I, Y^J, F_I^j), \delta X^I, 0, F_L^j \delta P_I^L),$$

satisfy Eq. (6.21) and $\delta X^I(X) = V_X$. Then, we have already proved that the grade of smooth uniformity of any point $X$ at $\mathcal{B}_1$ different from 0 is 2 and the smoothly uniform submanifolds are given by the spheres $S(\|X\|)$. Then, the grade of uniformity of 0 is 0 and the smoothly uniform submanifold at 0 is $\{0\}$ (exercise). Therefore, ignoring the origin point, $\mathcal{B}_1$ is a "laminated" body covered by smoothly uniform submanifolds of dimension 2 with a kind of structure similar to *liquid crystals*. The uniform-material foliation is again the same foliation (exercise: use a similar argument to Eq. (6.23)). Notice that the picture of this material is similar to the previous one (Fig. 6.1) but the "*solid core*" is just a point. However, in this case, the homogeneity is not ensured.

Let us now test the (local) *homogeneity* of $\mathcal{B}$. In this sense, by using again Proposition 6.2.13, we should study the existence of a system of (local) coordinates $(x^i)$ at each $X \in \mathcal{B}_1$ such that

$$\frac{\partial W}{\partial x^l} = 0, \tag{6.25}$$

for all $l \leq 2$ if $X \neq 0$ and $l = 0$ if $X = 0$.

Let $(x^i)$ be a system of local coordinates of $\mathcal{B}_1$. Using the chain rule we have that

$$\frac{\partial \mathcal{W}}{\partial x^i} = \frac{\partial \widehat{W}}{\partial r}\frac{\partial r}{\partial x^i} + \frac{\partial \widehat{W}}{\partial J}\frac{\partial J}{\partial x^i}.$$

Therefore, the immersion property of $\widehat{W}$ implies that $(x^i)$ are homogeneous coordinates if and only if

(1)\* $\frac{\partial r}{\partial x^l} = 0$,
(2)\* $\frac{\partial J}{\partial x^l} = 0$,

for all $l \leq 2$ if $X \neq 0$ and $l = 0$ if $X = 0$. Hence, the study of homogeneity depends only on the properties of $r$ and $J$.

For each $j^1_{X,Y}\phi \in \Pi^1(\mathcal{B}_1, \mathcal{B}_1)$

$$r\left(j^1_{X,Y}\phi\right) = g\left(Y\right)\left(T_X\phi\left(e\left(X\right)\right), T_X\phi\left(e\left(X\right)\right)\right) + \parallel X \parallel^2$$

$$= e^i\left(X\right)e^j\left(X\right)\frac{\partial\phi^k}{\partial x^i_{|X}}\frac{\partial\phi^l}{\partial x^j_{|X}}g_{kl}\left(Y\right) + \parallel X \parallel^2,$$

where, in this case, $e^j$ are the coordinates of $e$ respect to $(x^i)$. So, considering the induced coordinates $(x^i, y^j, y^j_i)$ of $(x^i)$ on $\Pi^1(\mathcal{B}_1, \mathcal{B}_1)$ we have that for all $(\tilde{X}, \tilde{Y}, \tilde{F})$,

$$r \circ (x^i, y^j, y^j_i)^{-1}(\tilde{X}, \tilde{Y}, \tilde{F})$$

$$= e^i\left(X\right)e^j\left(X\right)\tilde{F}^k_i\tilde{F}^l_j g_{kl}\left(Y\right) + \left(X^I\right)^2,$$

where $X = (x^i)^{-1}(\tilde{X})$ and $Y = (x^i)^{-1}(\tilde{Y})$. In this way,

$$\frac{\partial r}{\partial x^k_{|j^1_{X,Y}\phi}} = 2\frac{\partial e^i}{\partial x^k_{|X}}e^j\left(X\right)\tilde{F}^k_i\tilde{F}^l_j g_{kl}\left(Y\right) + 2X^L\frac{\partial X^L}{\partial x^k_{|X}}.$$

So, let us study the equation,

$$\frac{\partial r}{\partial x^k} = 0.$$

Again, the dependence of the matrix variable on the left side of the equations led us to the necessary equation,

$$X^l \frac{\partial X^L}{\partial x^k_{|X}} = 0.$$

Hence, by using the non-degeneracy of $g$ we have that $\frac{\partial r}{\partial x^k} = 0$ if, and only if,

**(i)**

$$X^L \frac{\partial X^L}{\partial x^k} = 0, \tag{6.26}$$

**(ii)**

$$\frac{\partial e^i}{\partial x^k} = 0, \ \forall i. \tag{6.27}$$

Thus, **(1)\*** is satisfied if and only if

$$\frac{\partial e^i}{\partial x^l} = 0, \quad X^r \frac{\partial X^r}{\partial x^l} = 0, \ \forall i, \ l \leq 2. \tag{6.28}$$

These two equations can be translated by stating that the functions $e^i$ are constant on the spheres and the partial derivatives $\frac{\partial}{\partial x^l}$ are tangent to the spheres for $l \leq 2$.

Next, let us study condition **(2)\***. Notice that,

$$[J \circ (x^i, y^j, y^j_i)^{-1}](\tilde{X}, \tilde{Y}, \tilde{F})$$

$$= J(j^1_{((x^i)^{-1}(\tilde{X}), (y^j)^{-1}(\tilde{Y}))}((y^j)^{-1} \circ \tilde{\phi} \circ (x^i)))$$

$$= J(\nabla_{\tilde{Y}} (y^j)^{-1}) J(\tilde{F}) J(\nabla_{(x^i)^{-1}(\tilde{X})} (x^i)),$$

where $\tilde{F} = \nabla_{\tilde{X}} \tilde{\phi}$ and $(\tilde{X}, \tilde{Y}, \tilde{F})$ is in the codomain of $(x^i, y^j, y^j_i)$. Then, $\frac{\partial J}{\partial x^i} = 0$ if and only if

$$\frac{\partial}{\partial X^I} (J(\nabla_{(x^i)^{-1}(\tilde{X})} (x^i))) = 0. \tag{6.29}$$

So, denoting $(x^i)^{-1}(\tilde{X})$ by $X$, Eq. (6.29) is equivalent to

$$\frac{\partial^2 x^k}{\partial X^I \partial X^M_{|X}} = 0, \quad \forall k, M.$$

Therefore, **(2)\*** is can be expressed as follows:

$$\frac{\partial^2 x^i}{\partial X^K \partial X^M} = 0, \quad \forall i, M, \forall K \leq 2. \tag{6.30}$$

We conclude that $\mathcal{B}_1$ is (locally) homogeneous if and only if there exists a local system of coordinates $(x^i)$ which satisfies that the functions $e^i = e(x^i)$ are constants on the spheres, the partial derivatives $\frac{\partial}{\partial x^i}$ are tangent to the spheres and it satisfies Eq. (6.30).

Therefore, in general, $\mathcal{B}_1$ is not (locally) homogeneous. In fact, let us consider the following vector field:

$$e = X^K \frac{\partial}{\partial X^K} + r \frac{\partial}{\partial X^L},$$

with $L \leq 2$. The factor $r \frac{\partial}{\partial X^L}$ is added to get that the vector field $e$ does not vanish.

Assume that there exists a local system of homogeneous coordinates $(x^i)$ on $\mathcal{B}_1$. Notice that,

$$e = \left(X^K\right) \frac{\partial x^i}{\partial X^K} \frac{\partial}{\partial x^i} + r \frac{\partial x^i}{\partial X^L} \frac{\partial}{\partial x^i},$$

i.e., the coordinates $e^i$ respect to the coordinates $(x^i)$ are given by $X^k \frac{\partial x^i}{\partial X^K} + r \frac{\partial x^i}{\partial X^L}$. Then, it should satisfy that for each $l \leq 2$

$$\begin{aligned}
\frac{\partial e^i}{\partial x^l} &= \frac{\partial}{\partial x^l} \left( X^K \frac{\partial x^i}{\partial X^K} \right) + r \frac{\partial^2 x^i}{\partial x^l \partial X^L} \\
&= \frac{\partial X^K}{\partial x^l} \frac{\partial x^i}{\partial X^k} + X^K \frac{\partial^2 x^i}{\partial x^l \partial X^K} + r \frac{\partial^2 x^i}{\partial x^l \partial X^L} \\
&= \delta^i_l + X^K \frac{\partial^2 x^i}{\partial x^l \partial X^K} + r \frac{\partial^2 x^i}{\partial x^l \partial X^L} = 0.
\end{aligned}$$

Notice that,

$$\begin{aligned}
\frac{\partial^2 x^i}{\partial x^l \partial X^L} &= \frac{\partial}{\partial X^L} \left( \frac{\partial x^i}{\partial X^L} \circ \left(x^i\right)^{-1} \right) \\
&= \frac{\partial^2 x^i}{\partial X^K \partial X^L} \frac{\partial X^K}{\partial x^l} = 0.
\end{aligned}$$

This is a consequence of Eq. (6.30). So,

$$\frac{\partial e^i}{\partial x^l} = 0,$$

if and only if

$$\delta^i_l = -X^K \frac{\partial^2 x^i}{\partial x^l \partial X^K}.$$

(6.31)

Observe that, Eq. (6.31) implies that for $i \neq l$,

$$X^K \frac{\partial^2 x^i}{\partial x^l \partial X^K} = 0.$$

Thus, considering $X = (0, 0, c)$ with $c \neq 0$ we have that

$$\frac{\partial^2 x^i}{\partial x^l \partial X^3_{|X}} = \frac{\partial^2 x^i}{\partial X^K \partial X^3_{|X}} \frac{\partial X^K}{\partial x^l_{|X}}$$

$$= \frac{\partial^2 x^i}{\partial X^3 \partial X^3_{|X}} \frac{\partial X^3}{\partial x^l_{|X}} = 0.$$

Then $\frac{\partial^2 x^i}{\partial X^3 \partial X^3_{|X}} = 0$ or $\frac{\partial X^3}{\partial x^l_{|X}} = 0$. Observe that, by Eq. (6.31), for each $l \leq 2$

$$-\frac{1}{c} = \frac{\partial^2 x^l}{\partial X^3 \partial X^3_{|X}} \frac{\partial X^3}{\partial x^l_{|X}}.$$

So, the expression $\frac{\partial X^3}{\partial x^l_{|X}}$ cannot be zero. For the same reason, for $i \leq 2$, $\frac{\partial^2 x^i}{\partial X^3 \partial X^3_{|X}}$ is different to 0. Therefore, the above equation cannot be satisfied and the laminated simple liquid crystal $\mathcal{B}_1$ induced by this vector field $e$ is not homogeneous.

# Appendix A

# Foliations and Distributions

This part of the appendix is devoted to give a brief introduction to *foliations* and *distributions*. All the results and definitions exposed here can be found in Michor (2008) and Dufour and Zung (2005) (see also Reinhart, 1983).

Intuitively speaking, a foliation of a manifold $M$ is a decomposition of $M$ into immersed submanifolds, called the leaves of the foliation, which, in some way explained below, fits together nicely. These leaves are not necessarily of the same dimension.

**Definition A.0.1.** A *smooth singular foliation* or simply *foliation* on a smooth manifold $M$ is a partition $\mathcal{F} := \{\mathcal{F}(x)\}$ of $M$ into a disjoint union of smooth immersed connected submanifolds $\mathcal{F}(x)$, called *leaves*, which satisfies the following local foliation property at each point $x \in M$: Denote the leaf that contains $x$ by $\mathcal{F}(x)$, the dimension of $\mathcal{F}(x)$ by $k$ and the dimension of $M$ by $n$.

Then, there is a smooth local chart of $M$ with coordinates $(y^1, \ldots, y^n)$ in a neighborhood $U$ of $x$, $U := \{-\epsilon < y^1 < \epsilon, \ldots, -\epsilon < y^n < \epsilon\}$, such that the $k$-dimensional disk $\{y^{k+1} = \cdots = y^n = 0\}$ coincides with the path-connected component of the intersection of $\mathcal{F}(x)$ with $U$ which contains $x$, and each $k$-dimensional disk $\{y^{k+1} = c_{k+1}, \ldots y^n = c_n\}$, where $c_{k+1}, \ldots, c_n$ are constants, is wholly contained in some leaf of $\mathcal{F}$.

This (local) chart is called a *foliation chart* of $\mathcal{F}$ at $x$. A *foliation atlas* of dimension $k$ is an atlas of $M$ given by foliation charts. The *dimension* of the foliation $\mathcal{F}$ is a map $\dim_{\mathcal{F}} : M \to \{0, 1, \ldots, n\}$ such that, for each $x \in M$, $\dim_{\mathcal{F}}(x)$ is the dimension of the leaf $\mathcal{F}(x)$ at $x$.

If all the leaves of a singular foliation $\mathcal{F}$ have the same dimension, then one says that $\mathcal{F}$ is a *regular foliation*. Furthermore, in this case the dimension of $\mathcal{F}$ is a constant map and, hence, is identified with a number $k$ equal to the dimension of the leaves of $\mathcal{F}$.

Let $M$ be a manifold and $\mathcal{F}$ be a foliation on $M$. Consider a foliation chart of $M$ at a point $x \in M$ with coordinates $(y^1, \ldots, y^n)$ in a neighborhood $U := \{-\epsilon < y^1 < \epsilon, \ldots, -\epsilon < y^n < \epsilon\}$ of $x$. Then, the subset $U_0 := \{y^{k+1} = \cdots = y^n = 0\}$ of $U$ given by the condition of that the last $n - k$ coordinates are 0 is an open subset of the leaf $\mathcal{F}(x)$ (with the induced topology) which contains $x$. Indeed, we may define a local chart of $\mathcal{F}(x)$ at $x$ over $U_0$ by restricting the map $(y^1, \ldots, y^k)$, where $k$ is the dimension of $\mathcal{F}(x)$, to $U_0$. If $\mathcal{F}$ is a regular foliation, the restriction of the map $(y^1, \ldots, y^k)$ to each of the subsets $U_{c_1, \ldots, c_k} := \{y^{k+1} = c_{k+1}, \ldots y^n = c_n\}$, with $c_i$ constant for all $i = k + 1, \ldots n$, defines a local chart of the leaf in which $U_{c_1, \ldots, c_k} := \{y^{k+1} = c_{k+1}, \ldots y^n = c_n\}$ is contained.

**Definition A.0.2.** Let $M$ be a smooth manifold and $\mathcal{F}$ be a foliation of $M$. The *space of leaves* $M/\mathcal{F}$ is the quotient space of $M$, obtained by identifying two points of $M$ if they lie on the same leaf of $\mathcal{F}$.

Now, we will define a category associated to the foliations of the smooth manifolds. Thus, we need define the objects and the morphisms.

**Definition A.0.3.** A *foliated manifold* is a pair $(M, \mathcal{F})$, where $M$ is a smooth manifold and $\mathcal{F}$ a foliation of $M$. If the foliation $\mathcal{F}$ is regular, the pair $(M, \mathcal{F})$ will be called *regular foliated manifold*.

A *morphism between foliated manifolds* $f : (M, \mathcal{F}) \to (N, \mathcal{S})$ is a smooth map $f : M \to N$ which maps leaves of $\mathcal{F}$ into the leaves of $\mathcal{S}$.

Thus, since it is obvious that the composition of morphisms between foliated manifolds is again a morphism between foliated manifolds, we may construct a category of foliated manifolds. We will denote this category by $\mathcal{FM}$.

Note that, in the particular case of a regular foliated manifold $(M, \mathcal{F})$, the foliation $\mathcal{F}$ determines a foliation atlas such that the change-of-charts diffeomorphisms of foliation charts $\varphi_i$, $\varphi_j$ are of the form

$$\varphi_i \circ \varphi_j^{-1}(x, y) = (g_{ij}(x, y), h_{ij}(y)), \tag{A.1}$$

with respect to the decomposition $\mathbb{R}^n = \mathbb{R}^k \times \mathbb{R}^{n-k}$.

Conversely, let $\{(U_i, \varphi_i)\}$ be an atlas of a manifold $M$ such that the change-of-charts diffeomorphisms of $\varphi_i$, $\varphi_j$ satisfy Eq. (A.1). Then, each $U_i$ is divided into *plaques*, which are the connected components of the submanifolds $\varphi_i^{-1}(\mathbb{R}^k \times \{y\})$, $y \in \mathbb{R}^{n-k}$, and the change-of-chart diffeomorphisms preserve this division. The plaques globally amalgamate into leaves. Thus, $\{(U_i, \varphi_i)\}$ determines a unique regular foliation.

So, we have proved the following result.

**Proposition A.0.4.** *Let $M$ be a manifold with an atlas $\{(U_i, \varphi_i)\}_i$. Then, $\{(U_i, \varphi_i)\}_i$ is a foliation atlas associated to a unique regular foliation $\mathcal{F}$ of $M$ if and only if the changes of coordinates satisfy Eq. (A.1).*

We can give another equivalent definition.

**Proposition A.0.5.** *A regular foliation $\mathcal{F}$ of a manifold $M$ can be equivalently described in the following way: An open cover $\{U_i\}$ of $M$ with submersions $s_i : U_i \to \mathbb{R}^{n-k}$ (where $k$ is the dimension of $\mathcal{F}$) such that there are diffeomorphisms (necessarily unique)*

$$\gamma_{ij} : s_j(U_i \cap U_j) \to s_i(U_i \cap U_j),$$

*with $\gamma_{ij} \circ s_{j|U_i \cap U_j} = s_{i|U_i \cap U_j}$.*

*Note that, by unicity, the diffeomorphisms $\gamma_{ij}$ satisfy the cocycle condition $\gamma_{ij} \circ \gamma_{jk} = \gamma_{ik}$.*

**Proof.** If $(U_i, s_i, \gamma_{ij})$ is a triple on $M$ satisfying the above condition, using the rank theorem, we can construct an atlas $\{(V_j, \varphi_j)\}_j$ so that each $V_j$ is a subset of an $U_{\alpha_j}$ and there exists a diffeomorphism $\psi_j : s_{\alpha_j}(V_j) \to \mathbb{R}^{n-k}$, such that

$$\psi_j \circ s_{\alpha_j} = pr_2 \circ \varphi_j,$$

where $pr_2 : \mathbb{R}^k \times \mathbb{R}^{n-k} \to \mathbb{R}^{n-k}$ is the projection onto the last $(n-k)$ components. This atlas is a foliation atlas. In fact, if $(x,y) \in \varphi_i(U_i \cap U_j) \subset \mathbb{R}^k \times \mathbb{R}^{n-k}$, then,

$$
\begin{aligned}
pr_2 \circ \varphi_j \circ \varphi_i^{-1}(x,y) &= \left(\psi_j \circ s_{\alpha_j} \circ \varphi_i^{-1}\right)(x,y) \\
&= \left(\psi_j \circ \gamma_{\alpha_j \alpha_i} \circ s_{\alpha_i} \circ \varphi_i^{-1}\right)(x,y) \\
&= \left(\psi_j \circ \gamma_{\alpha_j \alpha_i} \circ \psi_i^{-1}\right)(y).
\end{aligned}
$$

Conversely, if $\{(U_j, \varphi_j)\}$ is an atlas such that the change-of-chart diffeomorphisms are of the form of Eq. (A.1), we take $s_j = pr_2 \circ \varphi_j$ and $\gamma_{ij} = h_{ij}$ (see Eq. (A.1)). $\qquad \square$

A triple $(U_i, s_i, \gamma_{ij})$ satisfying conditions of Proposition A.0.5 is called *the Haefliger cocycle representing* $\mathcal{F}$.

**Example A.0.6.** The space $\mathbb{R}^n$ admits a (regular) foliation of dimension $k$, for which the foliation atlas consists of only one chart, the identity map $Id : \mathbb{R}^n \to \mathbb{R}^k \times \mathbb{R}^{n-k}$. Then, the leaves are

$$
\mathbb{S}(x_0, y_0) = \{(x, y_0) \in \mathbb{R}^k \times \mathbb{R}^{n-k} : x \in \mathbb{R}^k\},
$$

for each $(x_0, y_0) \in \mathbb{R}^k \times \mathbb{R}^{n-k}$. Generally, consider a map $f : \mathbb{R}^n \to \{0, 1, \dots, n\}$ such that if $f(x^1, \dots, x^n) = r$, it satisfies that

$$
f\left(y^1, \dots, y^k, x^{k+1}, \dots, x^n\right) = r.
$$

Then, there exists a unique foliation $\mathbb{S}$ of $\mathbb{R}^n$ with $f$ as the dimension map. In fact, the leaves of this foliation are given by

$$
\mathbb{S}(x^1, \dots, x^n) := \{(y^i, x^j) : y^i \in \mathbb{R}, \ i = 1, \dots, f(x^1, \dots, x^n)\},
$$

This foliation will be called *trivial or canonical foliation of* $\mathbb{R}^n$ *associated to* $f$.

Now, we will give a result which justify the name of *canonical foliation*.

**Proposition A.0.7.** *Let* $M$ *be a manifold and* $\mathcal{F}$ *be a partition into subsets of* $M$. *Denoting by* $\mathcal{F}(x)$ *the subset of the partition* $\mathcal{F}$ *which contains* $x \in M$, *we have that* $\mathcal{F}$ *is a foliation if and only if there exists a canonical foliation* $\mathbb{S}$ *of* $\mathbb{R}^n$ *such that we can cover* $M$ *by local* *charts of* $\varphi : U \to \mathbb{R}^n$ *satisfying that*

$$\mathbb{S}\left(\varphi\left(x\right)\right) \cap \varphi\left(U\right) = \varphi\left(\mathcal{F}\left(x\right) \cap U\right),$$

*for all* $y \in U$.

**Proof.** The proof follows from the definition. □

In a more categorical way we can summarize the above result as follows: *a partition* $\mathcal{F}$ *of a manifold* $M$ *is a foliation if and only if* $(M, \mathcal{F})$ *is locally isomorphic (in the categorical sense) to some canonical foliated manifold* $(\mathbb{R}^n, \mathbb{S})$.

**Example A.0.8.** Any submersion $\pi : M \to N$ defines a regular foliation of $M$ whose leaves are the connected components of the fibres of $\pi$. The dimension of the leaves is equal to the codimension of $N$. A foliation atlas is derived from the rank theorem. Foliations associated to the submersions are also called *simple foliations*. The foliations associated to submersions with connected fibres are called *strictly simple*.

Note that, it is easy to prove that a simple foliation is strictly simple precisely when its space of leaves is Hausdorff.

**Example A.0.9.** Let $(M, \mathcal{F})$ and $(N, \mathcal{S})$ be two foliated manifolds. Then there is the *product foliation* $\mathcal{F} \times \mathcal{S}$ on $M \times N$, whose leaves are the products of leaves of $\mathcal{F}$ and $\mathcal{S}$. Furthermore,

$$T\left(\mathcal{F} \times \mathcal{S}\right) = T\mathcal{F} \oplus T\mathcal{S}.$$

**Example A.0.10.** Let $f : N \to M$ be a smooth map and $\mathcal{F}$ be a regular foliation of $M$ of dimension $k$. Assume that $f$ is transversal to $\mathcal{F}$. Then we can get a foliation $f^{-1}\left(\mathcal{F}\right)$ on $N$ as follows.

Suppose that $\mathcal{F}$ is given by the Haefliger cocycle $(U_i, s_i, \gamma_{ij})$ on $M$. Put $V_i = f^{-1}\left(U_i\right)$ and $s_i' = s_i \circ f_{|V_i}$. Note that, for each $x \in N$ such that $f\left(x\right) \in U_i$

$$T_x s_i' = T_{f(x)} s_i \circ T_x f,$$

is surjective. In fact, $T_{f(x)} s_i$ is surjective and trivial on $T_{f(x)} \mathcal{F}$ (see the proof of the Proposition A.0.5). Thus, using that

$$T_x f\left(T_x N\right) + T_{f(x)} \mathcal{F} = T_{f(x)} M,$$

it is easy to prove that $T_x s_i'$ is surjective.

The foliation $f^{-1}(\mathcal{F})$ is now given by the Haefliger cocycle $(V_i, s'_i, \gamma_{ij})$ on $N$. We have that

$$\text{codim}\left(f^{-1}(F)\right) = \text{codim}(\mathcal{F}),$$

and,

$$T\left(f^{-1}(\mathcal{F})\right) = (Tf)^{-1}(T\mathcal{F}).$$

As a consequence, $N$ is foliated by connected components of $f^{-1}(\mathcal{F}(x))$, where $\mathcal{F}(x)$ are the leaves of $\mathcal{F}$.

**Example A.0.11.** Let $\phi : G \times M \to M$ be a smooth action of a Lie group $G$ on a smooth manifold $M$. We say that the action of $G$ on $M$ is *foliated* if $\dim(G_x)$ (where $G_x$ is the isotropy group on $x$) is a constant function of $x$. In this case, the connected components of the orbits of the action are the leaves of a regular foliation of $M$.

Let $(M, \mathcal{F})$ be a foliated manifold. Then, the tangent spaces of the leaves of the foliation defines a map which associates to any point $x$ of $M$ a subspace of the tangent space $T_x M$. So, we obtain what is called a *distribution on $M$* (see Example A.0.14).

**Definition A.0.12.** A *singular distribution* or simply *distribution* $D$ on a smooth manifold $M$ is the assignment to each point $x$ of $M$ a vector subspace $D_x$, called *fibre at $x$*, of the tangent space $T_x M$. If the dimension of $D_x$ is constant the distribution is called a *regular distribution*.

The distribution $D$ is called *smooth* if for any point $x$ of $M$ and any vector $v_x \in D_x$, there is a smooth vector field $\Theta$ defined in a neighborhood $U_x$ of $x$ which is tangent to the distribution, i.e.,

$$\Theta(y) \in D_y, \quad \forall y \in U_x,$$

and such that $\Theta(x) = v_x$.

Let $D$ be a regular smooth distribution on a manifold $M$. Fix a point $x$ at $M$ and consider a basis $\{v_x^i\}_i$ of the tangent space $T_x M$. Then, for each $v_x^i$ there exists a (local) vector field $X^i$ tangent to the distribution $D$ such that $\Theta^i(x) = v_x^i$. By using the inverse function theorem and shrinking (if it is necessary) the domain of the vector fields $\Theta^i$, we prove that a regular distribution $D$ is smooth if and only if $D$ is locally finitely generated (see Definition A.0.21).

We will define a morphism to construct a category. Let $D_1$ and $D_2$ be two different distributions over the manifolds $M_1$ and $M_2$, respectively. A *morphism of distributions* from $D_1$ to $D_2$ is a pair of maps $(\Phi, \phi)$, with $\Phi : D_1 \to D_2$ and $\phi : M_1 \to M_2$, commuting according to the diagram

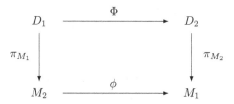

where $\pi_Q : TQ \to Q$ defines the canonical projection of the tangent bundle of a manifold $Q$. Thus,

$$\pi_{M_2} \circ \Phi = \phi \circ \pi_{M_1}.$$

This kind of morphism permits us to construct the category of distributions which will be denoted by $\mathcal{D}$. Obviously, $\phi$ is characterized by $\Phi$ and, hence, the morphism will be sometimes denoted by $\Phi$.

A morphism $(\Phi, \phi)$ from $D_1$ to $D_2$ is said to be *smooth* if for any vector field $\Theta$ tangent to $D_1$ it satisfies that $\Phi \circ \Theta$ is a differentiable map from $M_1$ to $TM_2$. Thus, the category of smooth distributions, denoted by $\mathcal{SD}$, consists of the smooth distributions and the smooth morphisms of distributions.

A natural notion associated to the concept of distribution is the following one.

**Definition A.0.13.** An *integral submanifold* of a distribution $D$ on a manifold $M$ is a connected immersed submanifold $N$ of $M$ such that for every $y \in N$ the tangent space $T_y N$ is a vector subspace of $D_y$. An integral submanifold $N$ is called *maximal* if it is not contained in any other integral submanifold and it is said to be of *maximal dimension* if its tangent space at every point $y$ is exactly $D_y$.

We say that a smooth distribution $D$ on a manifold $M$ is an *integrable distribution* if every point of $M$ is contained in a maximal integral submanifold of maximal dimension.

**Example A.0.14.** Let $(M, \mathcal{F})$ be a foliated manifold, then we may define the *tangent distribution associated to $\mathcal{F}$*, denoted by $D^{\mathcal{F}}$, given by the following correspondence

$$x \in M \mapsto D_x^{\mathcal{F}} = T_x \mathcal{F}(x).$$

It follows directly from the local foliation property that the tangent distribution is a smooth distribution.

Let $f : (M, \mathcal{F}) \to (N, \mathcal{S})$ be a morphism of foliated manifolds. Then, the restriction of the tangent map $Tf$ to $D^{\mathcal{F}}$ induces a smooth morphism of distributions between $D^{\mathcal{F}}$ and $D^{\mathcal{S}}$.

**Example A.0.15.** Let $C$ be a family of local smooth vector fields on $M$ such that its domains cover $M$. Then it gives rise to a smooth singular distribution $D^C$: for each point $x \in M$, $D_x^C$ is the vector space spanned by the values at $x$ of the vector fields of $C$ whose domains contain $x$. We say that $D^C$ is generated by $C$.

Also, associated to this example we have the following definition.

**Definition A.0.16.** A distribution $D$ is called *invariant with respect to a family of (local) smooth vector fields* $C$ if it is invariant with respect to every element of $C$: if $\Theta \in C$ and $\varphi_t^{\Theta} : U_t \to U_{-t}$ denotes the local flow of $\Theta$, then we have

$$T_x \varphi_t^{\Theta}(D_x) = D_{\varphi_t^{\Theta}(x)}, \quad \forall x \in M,$$

whenever $\varphi_t^{\Theta}$ is defined.

Finally, the following result, due to Stefan (1974) and Sussmann (1973), gives an answer to the following question: what are the conditions for a smooth singular distribution to be the tangent distribution of a singular foliation?

**Theorem A.0.17 (Stefan–Sussmann).** *Let $D$ be a smooth singular distribution on a smooth manifold $M$. Then the following three conditions are equivalent:*

(a) *$D$ is integrable.*
(b) *$D$ is generated by a family $C$ of smooth vector fields, and is invariant with respect to $C$.*
(c) *$D$ is the tangent distribution $D^{\mathcal{F}}$ of a smooth singular foliation $\mathcal{F}$.*

**Proof.** (a) ⇒ (b) Let $C$ be the family of all (local) vector fields which are tangent to $D$. The smoothness condition of $D$ implies that $D$ is generated by $C$. Furthermore, if $\Theta$ is an arbitrary vector field tangent to $D$, then $D$ is invariant with respect to $\Theta$. In fact, let $x$ be an arbitrary point in $M$, and denote by $\mathcal{F}(x)$ the maximal invariant submanifold of maximal dimension which contains $x$.

Then by definition (the condition of maximal dimension), for every point $y \in \mathcal{F}(x)$ we have

$$T_y \mathcal{F}(x) = D_y,$$

which implies that the vector field $\Theta$ restricted to $\mathcal{F}(x)$ is tangent to $\mathcal{F}(x)$, i.e., the restriction of $\Theta$ to $\mathcal{F}(x)$ gives rise to a vector field of $\mathcal{F}(x)$. In particular, the local flow $\varphi_t^\Theta$ of $\Theta$ can be restricted to $\mathcal{F}(x)$ (this fact follows from the maximality condition on $\mathcal{F}(x)$).

Finally, using that $\Theta$ is tangent to $\mathcal{F}(x)$ we have

$$T_x \varphi_t^\Theta(D_x) = T_x \varphi_t^\Theta(T_x \mathcal{F}(x)) = T_{\varphi_t^\Theta(x)} \mathcal{F}(x) = D_{\varphi_t^\Theta(x)}.$$

(b) ⇒ (c) Suppose that $D$ is generated by a family $C$ of smooth vector fields, and is invariant with respect to $C$. Let $x$ be an arbitrary point of $M$, denote by $k$ the dimension of $D_x$, and choose $k$ vector fields $\Theta_1, \ldots, \Theta_k$ of $C$ such that $\Theta_1(x), \ldots, \Theta_k(x)$ span $D_x$. Denote by $\varphi_t^{\Theta_1}, \ldots, \varphi_t^{\Theta_k}$ the local flow of $\Theta_1, \ldots, \Theta_k$, respectively. The map

$$(t_1, \ldots, t_k) \mapsto \varphi_{t_1}^{\Theta_1} \circ \cdots \circ \varphi_{t_k}^{\Theta_k}(x), \qquad (A.2)$$

is a local diffeomorphism from a $k$-dimensional disk to a $k$-dimensional submanifold containing $x$ in $M$. The invariance of $D$ with respect to $C$ implies that this submanifold is an integral submanifold of maximal dimension. Gluing these local integral submanifolds together (wherever they intersect), we obtain a partition of $M$ into a disjoint union of connected immersed integral submanifolds of maximal dimension, and thus, we may construct our foliation.

(c) ⇒ (a) If $D = D^{\mathcal{F}}$ is the tangent distribution of a singular foliation $\mathcal{F}$, then the leaves of $\mathcal{F}$ are maximal invariant submanifolds of maximal dimension for $D$. □

**Definition A.0.18.** An *involutive distribution* is a distribution $D$ such that if $\Theta_1, \Theta_2$ are two arbitrary vector fields which are tangent to $D$, then their Lie bracket $[\Theta_1, \Theta_2]$ is also tangent to $D$.

Using the Stefan–Sussmann theorem it is clear that if a singular distribution is integrable, then it is involutive. Conversely, for regular distributions we have the following theorem.

**Theorem A.0.19 (Frobenius).** *If a smooth regular distribution $D$ is involutive then it is integrable, i.e., it is the tangent distribution of a regular foliation.*

**Proof.** We only have to use Eq. (A.2) again to prove this result. □

Note that, if in Frobenius theorem we omit the word regular, then it is false. As a counterexample we may construct the following distribution.

**Example A.0.20.** Consider the following singular distribution $D$ on $\mathbb{R}^2$ given by

$$(x,y) \mapsto \begin{cases} T_{(x,y)}\mathbb{R}^2 & \text{if } x > 0, \\ \left\langle \dfrac{\partial}{\partial x_{|(x,y)}} \right\rangle & \text{if } x \leq 0. \end{cases}$$

Observe that $D$ does not have constant dimension ($\dim\left(D_{(x,y)}\right) = 2$ if $x > 0$ and $\dim\left(D_{(x,y)}\right) = 1$ if $x \leq 0$). Furthermore, it is trivial that $D$ is involutive.

Finally, for each $x > 0$, a maximal integral manifold of maximal dimension of $D$ which contains $(x,y)$, for all $y \in \mathbb{R}$, is given by

$$\mathbb{R}^+ := \{(x,y) \in \mathbb{R}^2 : x > 0\}.$$

Also, for each $(0,y_0)$, a maximal integral manifold of maximal dimension of $D$ which contains $(0,y_0)$, $L$, must verifies that

$$\dim(L) = 1, \quad T_{(0,y_0)}L = \left\langle \frac{\partial}{\partial x_{|(0,y_0)}} \right\rangle, \quad L \cap \mathbb{R}^+ = \emptyset.$$

Since that the point $(0,y_0)$ must belong to $L$, $L$ has to be of the form

$$L := \{(x,y_0) \in \mathbb{R}^2 : x \leq 0\},$$

which is not possible since the leaf of $D$ through a point is a manifold without boundary. Thus, $D$ is smooth involutive but not integrable.

Now, in order to repair this problem we are going to give the following definition.

**Definition A.0.21.** A smooth distribution $D$ on a manifold $M$ is called *locally finitely generated* if for any $x \in M$ there is a neighborhood $U$ of $x$ such that the $\mathcal{C}^\infty(U)$-module of smooth tangent vector fields to $D$ in $U$ is finitely generated: there is a finite number of smooth vector fields $X_1, \ldots, X_k$ on $U$ which are tangent to $D$, such that any smooth vector field $Y$ in $U$ which is tangent to $D$ can be written as

$$Y = f_i X_i,$$

with $f_i \in \mathcal{C}^\infty(U)$.

So, the situation in the finitely generated case is better:

**Theorem A.0.22 (Hermann).** *Any locally finitely generated smooth involutive distribution on a smooth manifold is integrable.*

See Abdelghani (2000) for more details.

# Appendix B

# Covariant Derivatives

This appendix is devoted to present a brief introduction to *covariant derivatives*. We will present important notions like the notion of *derivation* or *geodesic*. Furthermore, we will relate the concept of covariant derivative with the one introduced in Appendix A, the concept of distribution. We will also study the tensors of *torsion* and *curvature* relating them with the *Christoffel symbols*. For a more detailed survey see de León and Rodrigues (1989).

**Definition B.0.1.** Let $M$ be a manifold. A *derivation on $M$* is an $\mathbb{R}$-linear map $D : \mathfrak{X}(M) \to \mathfrak{X}(M)$ with a vector field $\Theta \in \mathfrak{X}(M)$ such that for each $f \in \mathcal{C}^\infty(M)$ and $\Xi \in \mathfrak{X}(M)$,

$$D(f\Xi) = fD(\Xi) + \Theta(f)\Xi.$$

We call $\Theta$ the *base vector field of $D$*. So, a derivation on $M$ is characterized by two geometrical objects, $D$ and $\Theta$.

A classical example of derivation is given by the bracket of vector fields on a manifold $M$. In fact, let $\Theta$ be a vector field on $M$, the operator given by fixing $\Theta$ in the Lie bracket

$$[\Theta, \cdot] : \mathfrak{X}(M) \to \mathfrak{X}(M),$$

is a derivation on $M$ with $\Theta$ as base vector field.

Another example comes from the so-called *covariant derivatives*.

**Definition B.0.2.** A *covariant derivative* or *linear connection* on $M$ is an $\mathbb{R}$-bilinear map $\nabla : \mathfrak{X}(M) \times \mathfrak{X}(M) \to \mathfrak{X}(M)$ such that,

(1) it is $\mathcal{C}^\infty(M)$-linear in the first variable;

(2) for all $\Theta, \Xi \in \mathfrak{X}(M)$ and $f \in \mathcal{C}^\infty(M)$,

$$\nabla_\Theta f\Xi = f\nabla_\Theta\Xi + \Theta(f)\Xi. \tag{B.1}$$

Then, given a covariant derivative $\nabla$, any vector field $\Theta \in \mathfrak{X}(M)$ generates a derivation on $M$, $\nabla_\Theta$, (with base vector field $\Theta$) fixing the first coordinate again, i.e.,

$$\nabla_\Theta : \mathfrak{X}(M) \to \mathfrak{X}(M),$$

such that

$$\nabla_\Theta(\Xi) = \nabla_\Theta\Xi, \quad \forall \Xi \in \mathfrak{X}(M).$$

**Remark B.0.3.** We will recover later (Appendix C) the notion of linear connection in a different but equivalent scenario, the linear frame bundle of the manifold.

**Remark B.0.4.** Let $\nabla$ be a covariant derivative on $M$. Consider four vector fields $\Theta_1, \Theta_2, \Xi_1, \Xi_2 \in \mathfrak{X}(M)$ and an open subset $U \subseteq M$ such that

$$\Theta_{1|U} = \Theta_{2|U}, \quad \Xi_{1|U} = \Xi_{2|U}.$$

Then,

$$[\nabla_{\Theta_1}\Xi_1]_{|U} = [\nabla_{\Theta_2}\Xi_2]_{|U}.$$

So, $\nabla$ may be naturally induced to a covariant derivative on $U$, i.e., a covariant derivative is of local type.

In fact, let $\Theta$ and $\Xi$ are other two vector fields on $M$ and $x \in M$ such that

$$\Theta(x) = \Xi(x).$$

Hence, due to the $\mathcal{C}^\infty(M)$-linearity, it satisfies that

$$\nabla_\Theta\Upsilon(x) = \nabla_\Xi\Upsilon(x), \quad \forall \Upsilon \in \mathfrak{X}(M).$$

Let $(x^i)$ be a local system of coordinates on an open subset $U$ of $M$. Then, there exist $m^3$ functions $\Gamma^k_{ij}$ on $U$ (with $m = \dim(M)$) such that

$$\nabla_{\frac{\partial}{\partial x^j}}\frac{\partial}{\partial x^i} = \Gamma^k_{ij}\frac{\partial}{\partial x^k}.$$

The local functions $\Gamma_{ij}^k$ are called *Christoffel symbols of* $\nabla$ *respect to* $\left(x^i\right)$. Let $\left(y^j\right)$ be another local system of coordinates on an open subset $V$ of $M$ such that $U \cap V \neq \emptyset$.

Then,

$$\frac{\partial}{\partial y^k} = \frac{\partial x^l}{\partial y^k} \frac{\partial}{\partial x^l},$$

on the intersection $U \cap V$. Denote the Christoffel symbols of $\nabla$ respect to $\left(y^j\right)$ by $\overline{\Gamma}_{ij}^k$. Therefore, the transformation rule for the Christoffel symbols is given by

$$\overline{\Gamma}_{ij}^k = \frac{\partial y^k}{\partial x^r} \frac{\partial x^l}{\partial y^j} \frac{\partial x^m}{\partial y^i} \Gamma_{ml}^r + \frac{\partial y^k}{\partial x^r} \frac{\partial^2 x^r}{\partial y^j \partial y^i}. \tag{B.2}$$

Let $\Theta$ and $\Xi$ be vector fields on $M$ such that the local expressions are given by

$$\Theta = \Theta^j \frac{\partial}{\partial x^j}, \quad \Xi = \Xi^i \frac{\partial}{\partial x^i}.$$

Then,

$$\nabla_\Theta \Xi = \left( \Gamma_{ij}^k \Theta^j \Xi^i + \Theta^j \frac{\partial \Xi^k}{\partial x^j} \right) \frac{\partial}{\partial x^k}. \tag{B.3}$$

**Definition B.0.5.** Let $\gamma : I \to M$ be a curve on $M$ and $\Theta$ be a vector field on $M$. We define the *covariant derivative* along $\gamma$ of $\Theta$ by

$$D\Theta/dt = \nabla_{\dot{\gamma}(t)} X.$$

We say that $\Theta$ *is parallel along* $\gamma$ if $D\Theta/dt = 0$. We say that $\gamma$ is a *geodesic of* $\nabla$ if $\dot{\gamma}(t)$ is parallel along $\gamma$, i.e.,

$$\nabla_{\dot{\gamma}(t)} \dot{\gamma}(t) = 0.$$

Notice that $D\Theta/dt$ is well-defined because of the dependence of $\nabla$ at the first variable is only on the value at the point (see Remark B.0.4). To work with the expression $\nabla_{\dot{\gamma}(t)} \dot{\gamma}(t)$, we will extend $\dot{\gamma}(t)$ by a local vector field. It is an easy exercise to prove that $\nabla_{\dot{\gamma}(t)} \dot{\gamma}(t)$ does not depend on the extension.

In local coordinates,

$$D\Theta/dt = \left( \frac{\partial \gamma^j}{\partial t} \Theta^i \Gamma^k_{ij} + \frac{\partial \left( \Theta^k \circ \gamma \right)}{\partial t} \right) \frac{\partial}{\partial x^k},$$

and

$$\nabla_{\dot{\gamma}(t)} \dot{\gamma}(t) = \left( \frac{\partial \gamma^j}{\partial t} \frac{\partial \gamma^i}{\partial t} \Gamma^k_{ij} + \frac{\partial^2 \gamma}{\partial t^2} \right) \frac{\partial}{\partial x^k},$$

where

$$\dot{\gamma}(t) = \frac{\partial \gamma^i}{\partial t} \frac{\partial}{\partial x^i}.$$

Therefore, $\gamma$ is a geodesic if it satisfies the following system of differential equations:

$$\frac{\partial^2 \gamma}{\partial t^2} + \Gamma^k_{ij} \frac{\partial \gamma^j}{\partial t} \frac{\partial \gamma^i}{\partial t} = 0, \quad \forall k. \tag{B.4}$$

So, from the existence theorem of ordinary differential equations, we easily deduce the following result.

**Theorem B.0.6.** *Let $\nabla$ be a covariant derivative on $M$. For any tangent vector $\Theta_x \in T_x M$ at a point $x \in M$, there is a unique geodesic $\gamma$ of $\nabla$ with initial conditions $\gamma(0) = x$ and $\dot{\gamma}(0) = \Theta_x$ defined on a local interval $(-\epsilon, \epsilon)$ for some $\epsilon > 0$.*

It is important to remark that the existence of geodesics is, in general, local.

**Definition B.0.7.** A covariant derivative $\nabla$ is *geodesically complete* if every geodesic may be extended so as to be defined for all $t \in \mathbb{R}$.

**Example B.0.8.** On $\mathbb{R}^m$ we may define the *canonical linear connection* on $\mathbb{R}^m$. Let $\left( X^I \right)$ be the canonical coordinates of $\mathbb{R}^m$. Then, the canonical linear connection $\nabla^m$ on $\mathbb{R}^m$ is characterized by

$$\nabla^m_{\frac{\partial}{\partial X^J}} \frac{\partial}{\partial X^I} = 0, \quad \forall I, J,$$

i.e., the Christoffel symbols of $\nabla^m$ respect to $\left( X^I \right)$ are zero. Then, the geodesic equation (B.4) becomes

$$\frac{\partial^2 \gamma}{\partial t^2} = 0, \quad \forall k.$$

Hence, the geodesics are the straight lines.

Associated to any covariant derivative $\nabla$ there are two important tensors:

- **Torsion:** $T(\Theta, \Xi) = \nabla_\Theta \Xi - \nabla_\Xi \Theta - [\Theta, \Xi]$, $\forall \Theta, \Xi \in \mathfrak{X}(M)$.
- **Curvature:** $R(\Theta, \Xi)\chi = \nabla_\Theta \nabla_\Xi \chi - \nabla_\Xi \nabla_\Theta \chi - \nabla_{[\Theta,\Xi]}\chi$, $\forall \Theta, \Xi, \chi \in \mathfrak{X}(M)$.

Notice that, clearly, torsion and curvature satisfy conditions of skew-symmetry, i.e.,

- for all $\Theta, \Xi \in \mathfrak{X}(M)$,

$$T(\Theta, \Xi) = -T(\Xi, \Theta);$$

- for all $\Theta, \Xi \in \mathfrak{X}(M)$,

$$R(\Theta, \Xi) = -R(\Xi, \Theta).$$

Let $(x^i)$ be a local system of coordinates on $M$. Then,

$$T\left(\frac{\partial}{\partial x^i}, \frac{\partial}{\partial x^j}\right) = \left(\Gamma^l_{ji} - \Gamma^l_{ij}\right)\frac{\partial}{\partial x^l} \tag{B.5}$$

and,

$$R\left(\frac{\partial}{\partial x^i}, \frac{\partial}{\partial x^j}\right)\frac{\partial}{\partial x^k} = \left(\Gamma^l_{kj}\Gamma^m_{li} - \Gamma^l_{ki}\Gamma^m_{lj} + \frac{\partial \Gamma^m_{kj}}{\partial x^i} - \frac{\partial \Gamma^m_{ki}}{\partial x^j}\right)\frac{\partial}{\partial x^m}. \tag{B.6}$$

For any three vector fields $\Theta, \Xi, \Upsilon$ we will sometimes denote $R(\Theta, \Xi)\Upsilon$ by $R(\Theta, \Xi, \Upsilon)$.

**Definition B.0.9.** Let $\nabla$ be a covariant derivative on $M$. $\nabla$ is said to be *symmetric* if its torsion tensor $T$ is zero. $\nabla$ is said to be *flat* is its curvature tensor $R$ is zero.

Notice that $\nabla$ is symmetric if and only if

$$\Gamma^k_{ij} = \Gamma^k_{ji} \tag{B.7}$$

for any system of local coordinates.

Let us consider a covariant derivative $\nabla$. Then, we may define $\nabla$ for tensor fields of type $(0, r)$ or type $(1, r)$.

Let $K$ be a $(0, r)$-type tensor on $M$ and a vector field $\Theta$ on $M$. Hence, $\nabla_\Theta K$ is a $(0, r)$-type tensor on $M$ such that

$$\nabla_\Theta K (\Theta_1, \ldots, \Theta_r) = \Theta (K (\Theta_1, \ldots, \Theta_r))$$
$$- \sum_l K (\Theta_1, \ldots, \nabla_\Theta \Theta_l, \ldots, \Theta_r). \quad \text{(B.8)}$$

On the other hand, assume that $K$ is a $(1, r)$-type tensor on $M$. Then, $\nabla_\Theta K$ is a $(1, r)$-type tensor on $M$ such that

$$\nabla_\Theta K (\Theta_1, \ldots, \Theta_r) = \nabla_\Theta (K (\Theta_1, \ldots, \Theta_r))$$
$$- \sum_l K (\Theta_1, \ldots, \nabla_\Theta \Theta_l, \ldots, \Theta_r). \quad \text{(B.9)}$$

In both cases, $\nabla_\Theta K$ is defined as the *covariant derivative of* $K$ *along* $\Theta$. So, we define $\nabla K$ as the $(0, r + 1)$-type tensor (respectively, $(1, r + 1)$-type tensor) given by

$$\nabla K (\Theta_1, \ldots, \Theta_r, \Theta) = \nabla_\Theta K (\Theta_1, \ldots, \Theta_r)$$

$\nabla K$ is called the *covariant derivative of* $K$. We say that $K$ is *parallel* if $\nabla K = 0$.

**Theorem B.0.10.** *Bianchi's identities Let $\nabla$ be a covariant derivative on $M$. Then, it satisfies that*

- **Bianchi's first identity**

$$\sum_{\text{cyc}} (R (\Theta, \Xi, \Upsilon)) = \sum_{\text{cyc}} [T (T (\Theta, \Xi), \Upsilon) + (\nabla_\Theta T) (\Xi, \Upsilon)];$$

- **Bianchi's second identity**

$$\sum_{\text{cyc}} [(\nabla_\Theta R) (\Xi, \Upsilon) + R (T (\Theta, \Xi), \Upsilon)] = 0,$$

*where $\sum_{\text{cyc}}$ denotes the cyclic sum with respect to $\Theta, \Xi$ and $\Upsilon$.*

**Proof.** The proof comes from using the expressions of torsion and curvature. For details see (Kobayashi and Nomizu, 1996). □

Notice that, in particular, for a symmetric covariant derivative it satisfies that

- for all $\Theta, \Xi, \Upsilon \in \mathfrak{X}(M)$

$$\sum_{\text{cyc}} (R(\Theta, \Xi, \Upsilon)) = 0;$$

- for all $\Theta, \Xi, \Upsilon \in \mathfrak{X}(M)$

$$\sum_{\text{cyc}} (\nabla_\Theta R)(\Xi, \Upsilon) = 0.$$

A very relevant example of covariant derivative comes from Riemannian geometry, the so-called *Levi-Civita connection*.

**Theorem B.0.11.** *Let $g$ be a Riemannian metric on $M$. Then, there exists a unique symmetric covariant derivative such that*

$$\nabla g = 0.$$

*This linear connection is called Levi-Civita connection. Furthermore, it satisfies the Koszul formula,*

$$g(\nabla_\Theta \Xi, \Upsilon) = \frac{1}{2}[\Theta(g(\Xi, \Upsilon)) + \Xi(g(\Upsilon, \Theta)) - \Upsilon(g(\Theta, \Xi))$$
$$- g([\Theta, \Upsilon], \Xi) - g([\Xi, \Theta], \Upsilon) - g([\Xi, \Upsilon], \Theta)].$$

**Proof.** Let us fix three vector fields $\Theta, \Xi, \Upsilon \in \mathfrak{X}(M)$. Then, by using $\nabla g = 0$ we have that

$$\nabla_\Theta g(\Xi, \Upsilon) = \Theta(g(\Xi, \Upsilon)) - g(\Xi, \nabla_\Theta \Upsilon) - g(\nabla_\Theta \Xi, \Upsilon) = 0. \quad \text{(B.10)}$$

Then,

$$\Theta(g(\Xi, \Upsilon)) + \Xi(g(\Upsilon, \Theta)) - \Upsilon(g(\Theta, \Xi))$$
$$= g(\Xi, \nabla_\Theta \Upsilon) + g(\nabla_\Theta \Xi, \Upsilon)$$
$$+ g(\Upsilon, \nabla_\Xi \Theta) + g(\nabla_\Xi \Upsilon, \Theta)$$
$$- g(\Theta, \nabla_\Upsilon \Xi) - g(\nabla_\Upsilon \Theta, \Xi)$$

On the other hand, the symmetry implies that

$$T(\Theta, \Xi) = \nabla_\Theta \Xi - \nabla_\Xi \Theta - [\Theta, \Xi] = 0.$$

Hence,

$$\Theta\left(g\left(\Xi,\Upsilon\right)\right) + \Xi\left(g\left(\Upsilon,\Theta\right)\right) - \Upsilon\left(g\left(\Theta,\Xi\right)\right)$$
$$= g\left(\Xi,\left[\Theta,\Upsilon\right]\right)$$
$$+ g\left(2\nabla_\Theta\Xi + \left[\Xi,\Theta\right],\Upsilon\right)$$
$$+ g\left(\left[\Xi,\Upsilon\right],\Theta\right),$$

The Koszul formula follows from this equation. Therefore, by using the non-degeneracy of $g$, we obtain the uniqueness and the existence of $\nabla$. To prove conditions (1) and (2) of the definition of covariant derivative we only have to use the Koszul formula and its non-degeneracy again. □

Let $\nabla$ be a covariant derivative on a manifold $M$. Then, we may construct the *opposite connection* $\tilde{\nabla}$ as the covariant derivative on a manifold $M$ given by

$$\tilde{\nabla}_\Theta\Xi = \nabla_\Xi\Theta + \left[\Theta,\Xi\right], \quad \forall\Theta,\Xi \in \mathfrak{X}\left(M\right).$$

Then, we have that the Christoffel symbols of $\tilde{\nabla}$ satisfy that,

$$\tilde{\Gamma}^k_{ij} = \Gamma^k_{ji}, \quad \forall i,j,k.$$

$\nabla$ is symmetric if and only if $\nabla = \tilde{\nabla}$.

Let $\left(x^i\right)$ be a local system of coordinates on an open subset $U$ of $M$. Then, we may construct the following local vector fields on $TM$

$$H_i = \frac{\partial}{\partial x^i} - v^j\Gamma^k_{ji}\frac{\partial}{\partial v^k}, \qquad \text{(B.11)}$$

where $\left(x^i,v^j\right)$ are the induced coordinates on $TU$. For all $v_x \in TU$, $\{H_i\left(v_x\right)\}_i$ is a linearly independent family of vectors at $T_{v_x}TM$. So, we define $H^\nabla_{v_x}$ as the vector subspace of $T_{v_x}TM$ spanned by $\{H_i\left(v_x\right)\}_i$. Then, covering $M$ by local charts, we construct a distribution $H^\nabla$ on $TM$ which is called *horizontal distribution of* $\nabla$. It is an exercise to prove that the definition of $H^\nabla$ does not depend on the choice of local charts.

Let $V$ be the *vertical distribution* of the tangent bundle $\pi_M :$ $TM \to M$, i.e.,

$$V_{v_x} = \mathrm{Ker}\,(T_{v_x}\pi_M)\,, \quad \forall v_x \in TM.$$

Then, it satisfies that

$$T\,(TM) = H^\nabla \oplus V, \tag{B.12}$$

i.e., $H^\nabla$ is a complementary distribution of the vertical distribution $V$.

Due to the decomposition (B.12), we may define the *horizontal projector* and the *vertical projector* as the following maps,

$$\cdot^h : T\,(TM) \to T\,(TM)\,, \quad \cdot^v : T\,(TM) \to T\,(TM)\,,$$

such that, for each $X_{v_x} \in T_{v_x}\,(TM)$

(i) $X_{v_x} = X^h_{v_x} + X^v_{v_x}$.
(ii) $X^h_{v_x} \in H^\nabla_{v_x}$.
(iii) $X^v_{v_x} \in V_{v_x}$.

Locally,

$$\left(\frac{\partial}{\partial x^i}\right)^h = \frac{\partial}{\partial x^i} - v^j \Gamma^k_{ji} \frac{\partial}{\partial v^k} = H_i,$$

$$\left(\frac{\partial}{\partial x^i}\right)^v = v^j \Gamma^k_{ji} \frac{\partial}{\partial v^k},$$

$$\left(\frac{\partial}{\partial v^i}\right)^h = 0,$$

$$\left(\frac{\partial}{\partial v^i}\right)^h = \frac{\partial}{\partial v^i}.$$

Then, for each vector field $\tilde{\Theta} = \Theta^i \frac{\partial}{\partial x^i} + \tilde{\Theta}^j \frac{\partial}{\partial v^j} \in \mathfrak{X}\,(TM)$, we have that

$$\tilde{\Theta}^h = \Theta^i \frac{\partial}{\partial x^i} - v^j \Theta^i \Gamma^k_{ji} \frac{\partial}{\partial v^k},$$

$$\tilde{\Theta}^v = v^j \Theta^i \Gamma^k_{ji} \frac{\partial}{\partial v^k} + \tilde{\Theta}^j \frac{\partial}{\partial v^j}.$$

On the other hand, by using again decomposition (B.12), the restriction of $T_{v_x}\pi_M$ to $H_{v_x}^\nabla$ is an isomorphism,

$$T_{v_x}\pi_M : H_{v_x}^\nabla \to T_x M.$$

Consequently, for any vector field $\Theta$ on $M$ there exists a unique vector field $\Theta^H$ on $TM$ tangent to $H^\nabla$ such that

$$T\pi_M \left(\Theta^H\right) = \Theta.$$

$\Theta^H$ is called the *horizontal lift* of $\Theta$. Obviously, for all $i$, we have that

$$\left(\frac{\partial}{\partial x^i}\right)^H = H_i, \tag{B.13}$$

and

$$\Theta^H = \Theta^i \frac{\partial}{\partial x^i} - v^j \Theta^i \Gamma_{ji}^k \frac{\partial}{\partial v^k}, \tag{B.14}$$

where $\Theta^i$ are the component of $\Theta$ respect to $\frac{\partial}{\partial x^i}$.

Reciprocally, let $H$ be a horizontal distribution on $TM$, i.e., a distribution on $TM$ satisfying that

$$T(TM) = H \oplus V.$$

Then, we may define the horizontal projector $\cdot^h$, the vertical projector $\cdot^v$ and the horizontal lift $\cdot^H$.

Notice that, for each $v_x \in T_x M$, the vertical lift is canonically defined as the following isomorphism:

$$\cdot_{v_x}^V : T_x M \to V_{v_x},$$

such that, for all $w_x \in T_x M$

$$(w_x)_{v_x}^V = (v_x + t w_x)'_{|0}.$$

Then, locally

$$\left(\frac{\partial}{\partial x^i}\right)^V = \frac{\partial}{\partial v^i}, \quad \forall i.$$

Denote the inverse of $\cdot_{v_x}^V$ by $\phi_{v_x}$.

Thus, we will define the *covariant derivative* $\nabla^H$ associated to $H$: for each $\Theta$ and $\Xi$ vector fields on $M$, $\nabla^H_\Theta \Xi$ is given by the following vector field:

$$\nabla^H_\Theta \Xi (x) = \phi_{\Xi(x)} ([T_x \Xi (\Theta (x))]^v), \qquad (B.15)$$

for all $x \in M$. It is easy to prove that, for a given covariant derivative $\nabla$ on $M$ we have

$$(\nabla)^{H^\nabla} = \nabla.$$

In fact, for each $i, j$

$$(\nabla)^{H^\nabla}_{\frac{\partial}{\partial x^j}} \frac{\partial}{\partial x^i} (x) = \phi_{\frac{\partial}{\partial x^i}_{|x}} \left( \left[ T_x \frac{\partial}{\partial x^i} \left( \frac{\partial}{\partial x^j}_{|x} \right) \right]^v \right)$$

$$= \phi_{\frac{\partial}{\partial x^i}_{|x}} \left( \Gamma^k_{ji} \frac{\partial}{\partial v^k} \left( \frac{\partial}{\partial x^i}_{|x} \right) \right)$$

$$= \Gamma^k_{ji} \frac{\partial}{\partial x^k}_{|x}$$

$$= \nabla_{\frac{\partial}{\partial x^j}} \frac{\partial}{\partial x^i} (x).$$

The converse is analogous.

**Theorem B.0.12.** *There exists a one-to-one correspondence between covariant derivatives and horizontal distributions.*

Let $\nabla$ be a covariant derivative on $M$. Assume that there is a local system of coordinates $(x^i)$ on $M$ such that

$$\nabla_{\frac{\partial}{\partial x^j}} \frac{\partial}{\partial x^i} = 0,$$

i.e., the Chistoffel symbols are zero. Then, obviously, $\nabla$ is symmetric and flat. In fact, we may prove the following result.

**Lemma B.0.13.** *Let $\nabla$ be a covariant derivative on a manifold $M$. $\nabla$ is flat and symmetric if and only if there exists an atlas $(x^i)$ of*

*M such that*

$$\nabla_{\frac{\partial}{\partial x^j}} \frac{\partial}{\partial x^i} = 0.$$

**Proof.** Let $(x^i)$ be a local system of coordinates. We will study the existence of (local) vector fields $\Theta$ such that

$$\nabla \Theta = 0.$$

So, $\Theta = \Theta^i \frac{\partial}{\partial x^i}$ should satisfy the following system of linear differential equations:

$$\frac{\partial \Theta^k}{\partial x^j} = \Gamma_{ij}^k \Theta^i, \quad j,k = 1, \ldots, m. \tag{B.16}$$

For a fixed $j$, we will denote the matrix $\Gamma_{ij}^k$ by $A^j$. Then, Eq. (B.16) turns into the following:

$$\frac{\partial \Theta}{\partial x^j} = A^j \Theta, \quad j = 1, \ldots, m, \tag{B.17}$$

where we are expressing $\Theta$ as the vector $(\Theta^1, \ldots, \Theta^m)$.

It is not complicated to verify the existence of uniqueness of solutions of this system for any initial condition. Hence, we may find a (local) basis of vector fields $\{\Theta_i\}$ such that

$$\nabla \Theta_i = 0, \quad \forall i.$$

On the other hand, due to the connection is symmetric, we have that

$$0 = T(\Theta_i, \Theta_j) = \nabla_{\Theta_j} \Theta_i - \nabla_{\Theta_i} \Theta_j - [\Theta_i, \Theta_j] = -[\Theta_i, \Theta_j], \quad \forall i, j.$$

Then, the basis $\{\Theta_i\}$ is composed by commuting vector fields and, hence, there exists a local system of coordinates $(y^j)$ such that

$$\Theta^j = \frac{\partial}{\partial y^j}, \quad \forall j.$$

$\square$

Note that, roughly speaking, we have proved that any flat and symmetric connection is (locally) a canonical linear connection (see Example B.0.8). In particular, respect to some local system of coordinates, the geodesics are the straight lines.

# Appendix C

# Principal Bundles and Connections

In this appendix, we will review of the most important results and definitions related with the notion of *principal bundles*. Convenient sources for a more complete exposition of principal bundles are Kobayashi and Nomizu (1996) and Sternberg (1983). Special attention is paid to the notions of *frame bundles* and *integrability* of its reduced subbundles. In 1950, C. Ehresmann (see Ehresmann, 1953, 1954, 1955a,b) formalized the notion of principal bundles and studied many $G$-structures associated in a natural way to an arbitrary manifold. We also remit to Bernard (1960), Chern (1966), Cordero *et al.* (1989) and Fujimoto (1972) for a detailed study on these topics.

## C.1  Principal Bundles

First, we will introduce the main notion in this appendix: *principal bundle*.

**Definition C.1.1.** Let $P$ be a manifold and $G$ be a Lie group which acts over $P$ by the right satisfying the following conditions:

(i) The action of $G$ is *free*, i.e.,

$$x \cdot g = x \Leftrightarrow g = e,$$

where $e \in G$ is the identity of $G$.

(ii) The canonical projection $\rho : P \to M = P/G$, where $P/G$ is the space of orbits, is a surjective submersion.

(iii) $P$ is *locally trivial*, i.e., $P$ is locally a product $U \times G$, where $U$ is an open set of $M$. More precisely, there exists a diffeomorphism $\Phi : \rho^{-1}(U) \to U \times G$, such that $\Phi(u) = (\rho(u), \phi(u))$, where the map $\phi : \rho^{-1}(U) \to G$ satisfies that

$$\phi(u \cdot g) = \phi(u) \cdot g, \quad \forall u \in U, \ \forall g \in G.$$

$\Phi$ is called a *trivialization on $U$*.

A principal bundle may be analogously defined by a left action. Along the book, we will not distinguish between a principal defined by right action and a principal bundle defined by a left action.

A principal bundle will be denoted by $P(M, G)$, or simply $\rho : P \to M$ if there is no ambiguity as to the structure group $G$. $P$ is called the *total space*, $M$ is the *base space*, $G$ is the *structure group* and $\rho$ is the *projection*. The closed submanifold $\rho^{-1}(x)$, $x \in M$ will be called the *fibre over $x$*. For each point $u \in P$, we have $\rho^{-1}(x) \triangleq uG$, where $\rho(u) = x$, and $uG$ will be called the *fibre through $u$*. Every fibre is diffeomorphic to $G$, but this diffeomorphism depends on the choice of the trivialization.

Now, we want to define the morphisms of this category.

**Definition C.1.2.** Given $P(M, G)$ and $P'(M', G')$ principal bundles, a principal bundle morphism from $P(M, G)$ to $P'(M', G')$ consists of a differentiable map $\Phi : P \to P'$ and a Lie group homomorphism $\varphi : G \to G'$ such that

$$\Phi(x \cdot g) = \Phi(x) \cdot \varphi(g), \quad \forall x \in P, \ \forall g \in G.$$

In this case, $\Phi$ maps fibres into fibres and it induces a differentiable map $\phi : M \to M'$ by the equality $\phi(x) = \rho(\Phi(u))$, where $u \in \rho^{-1}(x)$.

If these maps are embeddings, the principal bundle morphism will be called *embedding*. In such a case, we can identify $P$ with $\Phi(P)$, $G$ with $\varphi(G)$ and $M$ with $\phi(M)$ and $P(M, G)$ is said to be a *subbundle* of $P'(M', G')$. Furthermore, if $M = M'$ and $\varphi = Id_M$, $P(M, G)$ is called a *reduced subbundle* and we also say that $G'$ *reduces* to the subgroup $G$.

As usual, a principal bundle morphism is called *isomorphism* if it can be inverted by another principal bundle morphism. It is obvious that the property of being a principal bundle morphism is preserved by compositions and, indeed, principal bundles jointly with principal bundle morphisms give rise to a category denoted by $\mathcal{PB}$.

**Example C.1.3.** Let $M$ be an $n$-dimensional manifold and $G$ be a Lie group, then we can consider $M \times G$ as a principal bundle over $M$ with projection $pr_1 : M \times G \to M$ and structure group $G$. The action considered here is given by,

$$(x, g)\, h = (x, gh)\,, \quad \forall x \in M,\ \forall g, h \in G.$$

This principal bundle is called a *trivial principal bundle*. Equivalently, $G \times M$ may be considered as a principal bundle, the group $G$ acting on the left.

Using this example we can rewrite the condition of locally trivial: In the conditions of Definition C.1.1 $P$ is locally trivial if and only if it is locally isomorphic (in the sense of principal bundles) to the trivial principal bundle $pr_1 : M \times G \to M$ with $\varphi$ equal to the identity on $G$ (see Definition C.1.2).

Note the strong resemblance to the notion of trivial Lie groupoid in Example 3.0.7. This fact could gives us a clue about the close relation between transitive Lie groupoids and principal bundles.

Now, we will introduce an important example of principal bundle, the *frame bundle* of a manifold. In order to do that, we will start presenting the following definition.

**Definition C.1.4.** Let $M$ be an $n$-dimensional manifold. For each point $x \in M$ an ordered basis of $T_x M$ is called a *linear frame* at $x$.

**Remark C.1.5.** Alternatively, a linear frame at $x$ can be viewed as a linear isomorphism $\overline{x} : \mathbb{R}^n \to T_x M$ identifying a basis on $T_x M$ as the image of the canonical basis of $\mathbb{R}^n$ by $\overline{x}$.

We may use theory of jets to give a third way to interpret a linear frame. Let us give a very brief introduction to the notion of 1-jets of differentiable maps.

Let $\mathcal{C}^{\infty}(M, N)$ be the space of differentiable maps from $M$ to $N$ and fix $x \in M$. Then, we may define an equivalence relation

$\sim_x$ on $\mathcal{C}^\infty(M, N)$ in the following way: For each two maps $f, g \in \mathcal{C}^\infty(M, N)$,

$$f \sim_x g \Leftrightarrow f(x) = g(x) \quad \text{and} \quad T_x f = T_x g,$$

where $T_x f, T_x g : T_x M \to T_{f(x)} N$ are the induced maps on tangent spaces at $x$ of $f$ and $g$ respectively. The equivalence class of $f$ respect to $\sim_x$ is called 1-*jet of* $f$ *at* $x$ and is denoted $j_{x,y}^1 f$ with $y = f(x)$.

The proof of the following characterization is obvious.

**Lemma C.1.6.** *Let us consider two differentiable maps* $f, g : M \to N$ *such that, for a fixed* $x \in M$, $f(x) = g(x)$. *Then,* $j_{x,y}^1 f = j_{x,y}^1 g$ *if and only if for all* $i, j$

$$\frac{\partial\left(y^j \circ f\right)}{\partial x_{|x}^i} = \frac{\left(\partial y^j \circ g\right)}{\partial x_{|x}^i},$$

*for some local coordinates* $\left(x^i\right)$ *and* $\left(y^j\right)$ *on* $M$ *and* $N$, *respectively.*

Let $j_{x,y}^1 f$ and $j_{y,z}^1 g$ be two 1-jets of the differentiable maps $f : M \to N$ and $g : N \to S$ at the points $x \in M$ and $y \in N$. Then, we define the *composition* $\cdot$ *of* $j_{x,y}^1 f$ *and* $j_{y,z}^1 g$ as the 1-jet of the composition $g \circ f$ at $x$, i.e.,

$$j_{y,z}^1 g \cdot j_{x,y}^1 f = j_{x,z}^1 \left(g \circ f\right).$$

Now, a linear frame $\overline{x} : \mathbb{R}^n \to T_x M$ at $x \in M$ (see Remark C.1.5) may be considered as a 1-jet $j_{0,x}^1 \phi$ at $x$ of a local diffeomorphism $\phi$ from an open neighborhood of $0$ in $\mathbb{R}^n$ onto an open neighborhood of $x$ in $M$ such that $\phi(0) = x$ by imposing that $T_0 \phi = \overline{x}$. Here we are identifying $T_0 \mathbb{R}^n$ with $\mathbb{R}^n$ via the canonical isomorphism.

Thus, we denote by $FM$ the set of all linear frames at all the points of $M$. We can view $FM$ as a principal bundle over $M$ with structure group $Gl(n, \mathbb{R})$ and projection $\rho_M : FM \to M$ given by

$$\rho_M \left(j_{0,x}^1 \phi\right) = x, \quad \forall j_{0,x}^1 \phi \in FM.$$

Notice that any $g \in Gl(n, \mathbb{R})$ may be canonically identified by a 1-jet $j_{0,0}^1 F$ of an isomorphism $F$ from $0$ to $0$. So, the right action associated to this principal bundle of $Gl(n, \mathbb{R})$ over $FM$ is given by the composition of 1-jets.

This principal bundle is called *linear frame bundle* or simply *frame bundle of* $M$. Let $(x^i)$ be a local coordinate system on an open set $U \subseteq M$. Then we can introduce local coordinates $(x^i, x^i_j)$ over $FU \subseteq FM$ such that

- $x^i \left( j^1_{0,x} \phi \right) = x^i (x),$
- $y^j_i \left( j^1_{0,x} \phi \right) = \frac{\partial x^j \circ \phi}{\partial r^i_{|0}},$

for all $j^1_{0,x} \phi \in FU$ with $(r^i)$ the natural canonical (global) coordinates on $\mathbb{R}^n$. Notice that, by using these coordinates it is straightforward to prove that $\rho_M$ is a surjective submersion. In fact, $(x^i, x^i_j)$ induces the local trivialization from $FU$ to $U \times Gl(n, \mathbb{R})$.

Notice that this coordinates endows to the space $\Pi^1(\mathbb{R}^n, M)$ of all 1-jets of all differentiable maps $f : \mathbb{R}^n \to M$ at the point 0 with $f(0) = x$ of a differentiable structure of manifold such that $FM$ is an open subset.

Let $\psi : M \to N$ be a diffeomorphism from $M$ to $N$. Then, we can defined the *first prolongation of* $\psi$ as the isomorphism $F\psi : FM \to FN$ of principal bundles over $\psi$ given by

$$F\psi \left( j^1_{0,x} \phi \right) = j^1_{0,y} (\psi \circ \phi), \qquad (C.1)$$

for all $j^1_{0,x} \phi \in FM$ with $y = \psi(x)$. Notice that $F\psi$ is right invariant, i.e., for all $g \in Gl(n, \mathbb{R})$ we have that

$$F\psi \left( j^1_{0,x} \phi \cdot g \right) = F\psi \left( j^1_{0,x} \phi \right) \cdot g,$$

for all $j^1_{0,x} \phi \in FN$. It is also remarkable that the inverse of $F\psi$ is given by the first prolongation of the inverse of $\psi$; $(F\psi)^{-1} = F\psi^{-1}$. We denote by $e_1$ the frame $j^1_{0,0} Id_{\mathbb{R}^n} \in F\mathbb{R}^n$, where $Id_{\mathbb{R}^n}$ is the identity map on $\mathbb{R}^n$.

## C.2 G-structures

Next, as particular cases of reduced subbundles we shall introduce the notion of *G-structure*.

**Definition C.2.1.** A *G-structure over* $M$, $\omega_G(M)$, is a reduced subbundle of $FM$ with structure group $G$, a Lie subgroup of $Gl(n, \mathbb{R})$.

Now, we shall introduce the notion of *integrability of a G-structure*.

Note that there exists a principal bundle isomorphism $l : F\mathbb{R}^n \to \mathbb{R}^n \times Gl(n, \mathbb{R})$ over the identity map such that

$$l\left(j_{0,x}^1\phi\right) = \left(x, \left(J\phi_{|0}\right)\right), \quad \forall j_{0,x}^1\phi \in F\mathbb{R}^n,$$

where $J\phi_{|0}$ is the Jacobian matrix of $\phi$ at $0$. Indeed, we can consider the global section,

$$s : \mathbb{R}^n \to F\mathbb{R}^n,$$

$$x \mapsto j_{0,x}^1\tau_x,$$

where $\tau_x$ denote the translation on $\mathbb{R}^n$ by the vector $x$. So, a 1-jet $j_{0,x}^1\phi$ can be written in a unique way as a composition of $s(x)$ and a matrix of $Gl(n, \mathbb{R})$.

We have thus obtained a principal bundle isomorphism $F\mathbb{R}^n \cong \mathbb{R}^n \times Gl(n, \mathbb{R})$ over the identity map on $\mathbb{R}^n$. Then, if $G$ is a Lie subgroup of $Gl(n, \mathbb{R})$, we can transport $\mathbb{R}^n \times G$ by this isomorphism to obtain a $G$-structure on $\mathbb{R}^n$. This kind of $G$-structure will be called *standard flat* on $\mathbb{R}^n$.

**Definition C.2.2.** Let $\omega_G(M)$ be a $G$-structure over $M$. $\omega_G(M)$ is said to be *integrable* if we can cover $M$ by local charts $(U, \varphi_U)$ such that the restriction of the maps $l_{|F\varphi_U(U)} \circ F\varphi_U^{-1}$ to $\omega_G(M)$ is isomorphisms onto the trivial principal bundle $\varphi_U(U) \times G$ for some Lie group $G$.

Particularly, $\omega_G(M)$ is integrable if and only if for all point $x \in M$ there exists a local chart $(U, \varphi_U)$ through $x$ such that for all $j_{0,y}^1\psi \in \omega_G(M)$ with $y \in U$

$$j_{0,0}^1\left(\tau_{-\varphi_U(y)} \circ \varphi_U \circ \psi\right) \in G, \tag{C.2}$$

where $\tau_{-\varphi_U(y)}$ denote the translation on $\mathbb{R}^n$ by the vector $-\varphi_U(y)$.

Any $\{e_1\}$-structure on $M$, with $e_1$ the identity of $Gl(n, \mathbb{R})$, will be called *parallelism of* $M$. It is easy to show that any parallelism is, indeed, a global section of $\rho_M : FM \to M$. So, we will speak about *integrable sections*. Notice that, any parallelism $P : M \to FM$ is

locally written,
$$P\left(x^i\right) = \left(x^i, P_j^i\right).$$
On the other hand, using Eq. (C.2) we have that (locally) any integrable sections can be locally written as follows:
$$P_{|U} = j_{0,x}^1 \left(\varphi_U^{-1} \circ \tau_{\varphi_U(x)}\right),$$
with $(U, \varphi_U)$ a local chart on $M$. So, denoting $\varphi_U$ by $\left(x^i\right)$ and using the induced coordinates $(x^i, x_j^i)$, we can write that
$$P\left(x^i\right) = \left(x^i, \delta_j^i\right).$$
So, by using right translations, we can easily prove the following result.

**Proposition C.2.3.** *A $G$-structure $\omega_G(M)$ on $M$ is integrable if and only if for each point $x \in M$ there exists a local coordinate system $\left(x^i\right)$ on $M$ such that the local section,*

$$P\left(x^i\right) = \left(x^i, \delta_j^i\right), \tag{C.3}$$

*takes values into $\omega_G(M)$.*

Notice that this results guarantees the existence of local sections for integrable $G$-structures, but in general we cannot find global sections, which is a matter depending of the topological properties of the manifold. This result about integrability of $G$-structures may be found in Fujimoto (1972).

## C.3 Connections

We will now introduce the notion of *connections in principal bundles*. This concept will be closely related with the notion of linear connection (Appendix B).

Assume that $P(M, G)$ is a principal bundle with canonical projection $\rho : P \longrightarrow M$ and structure group $G$. We denote by $(V\rho)_u$ the vertical subspace at $u \in P$ with respect to the canonical projection, that is,

$$(V\rho)_u = \operatorname{Ker} T\rho(u).$$

**Definition C.3.1.** A connection $\Gamma$ in $P$ is an assignment of a subspace $H_u$ for each $u \in P$ such that

(1) $T_uP = (V\rho)_u \oplus H_u,$
(2) $H_{ug} = T(R_g)(H_u), \forall u \in P, \forall g \in G,$
(3) $H$ depends smoothly on $u$, that is, $H$ is a smooth distribution.

$H_u$ is called the horizontal subspace at $u$, and $H$ is the horizontal distribution.

Notice the similarity with the definition of the horizontal distribution associated to a covariant derivative (Eq. (B.12)). So, analogously, a tangent vector $X \in T_uP$ can be decomposed in its vertical and horizontal parts:

$$X = X_u^v + X_u^h. \tag{C.4}$$

Given a connection $\Gamma$ in $P$, we can define a 1-form $\omega$ on $P$ with values in the Lie algebra $\mathfrak{g}$ of $G$ as follows.

Since the action of $G$ on $P$, we know that every element $\xi \in \mathfrak{g}$ induces a vector field $\xi_P$ on $P$, the infinitesimal generator (see Remark 4.0.42), such that the map

$$\mathfrak{g} \longrightarrow (V\rho)_u,$$

$$\xi \mapsto \xi_P(u)$$

is a linear isomorphism. The, given $X_u \in T_uP$, $\omega(X_u)$ is the unique element $\xi \in \mathfrak{g}$ such that

$$\xi_P(u) = X_u^v.$$

$\omega$ is called the connection form of $\Gamma$, and it has the following properties:

(1) $\omega(X^h)0 = 0, \forall X \in TP;$
(2) $\omega(\xi_P) = \xi, \forall \xi \in \mathfrak{g};$
(3) $R_g^*\omega = Ad_{g^{-1}}\omega, \forall g \in G.$

Obviously, if $\omega$ is a 1-form on $P$ with values in $\mathfrak{g}$ and satisfying (ii) and (iii) of the above conditions, then the distribution

$$u \mapsto H_u = \ker \omega_u$$

defines a connection in $P$ with connection form $\omega$.

On the other hand, the linear mapping

$$T\rho(u) : H_u \longrightarrow T_x M,$$

where $x = \rho(u)$, is a linear isomorphism. So, given a vector field $Z$ on $M$, there exists a unique horizontal vector field $Z^H$ on $P$ such that

$$T\rho(Z^H) = Z,$$

$Z^H$ is called the horizontal lift of $Z$ to $P$ (with respect to the connection $\Gamma$).

### C.3.0.1 *Local expressions*

Assume that $(U_\alpha, x_\alpha^i)$ is an open neighborhood covering of $M$ such that we have trivializations

$$\Psi_\alpha : \rho^{-1}(U_\alpha) \longrightarrow U_\alpha \times G$$

with transition functions

$$\Psi_{\alpha\beta} : U_\alpha \cap U_\beta \longrightarrow G,$$

$$x \mapsto \Psi_{\alpha\beta}(x) = \Psi_\alpha(u)(\Psi_\beta(u))^{-1}.$$

Let $\sigma_\alpha ; U_\alpha \longrightarrow \rho^{-1}(U_\alpha)$ be the section

$$\sigma_\alpha(x) = \Psi_\alpha^{-1}(x, e), \quad \forall x \in U_\alpha.$$

On $U_\alpha \cap U_\beta$ we have

$$\theta_{\alpha\beta} = \Psi_{\alpha\beta}^* \theta,$$

where $\theta$ is the canonical Maurer–Cartan form on $G$. If $\omega$ is a connection form on $P$ and we put

$$\omega_\alpha = \sigma_\alpha^* \omega,$$

then we deduce that

$$\omega_\beta = Ad_{\Psi_{\alpha\beta}^{-1}} \omega_\beta + \theta_{\alpha\beta}.$$

**Remark C.3.2.** Recall that the Maurer–Cartan form $\theta$ on a Lie group $G$ is defined as follows:

$$\theta_g(v) = Tl_{g^{-1}}(v) \in \mathfrak{g}$$

for any $v \in T_g G$.

### C.3.0.2 *Curvature*

Let $P(M, G)$ be a principal bundle with a connection $\Gamma$, and let $\omega$ be the corresponding connection form. We define the *curvature* of $\Gamma$ as the 2-form

$$\Omega = (d\omega) \circ h,$$

where the map $h$ is given by the horizontal projection (C.4), that is

$$\Omega(X, Y) = d\omega(X^h, Y^h).$$

$\Omega$ is a 2-form with values in $\mathfrak{g}$ satisfying

$$R_g^* \Omega = Ad_{g^{-1}} \Omega.$$

Moreover, we have the following structure equation

$$d\omega = -\frac{1}{2}[\omega, \omega] + \Omega$$

that is,

$$d\omega(X, Y) = -\frac{1}{2}[\omega(X), \omega(Y)] + \Omega(X, Y), \quad \forall X, Y \in T_u P, \ \forall u \in P.$$

The *Bianchi's identity* states that

$$d\Omega \circ h = 0.$$

### C.3.0.3 *Parallel transport*

Let $\Gamma$ be a connection in a principal bundle $P(M, G)$. Given a curve $t \in [a, b] \mapsto \sigma(t) \in M$, a horizontal lift is a curve $\tilde{\sigma}$ in $P$ such that $\rho \circ \tilde{\sigma} = \sigma$ and $\dot{\tilde{\sigma}}(t)$ is a horizontal tangent vector for any $t \in [a, b]$.

A main result in Differential Geometry is the existence of horizontal lifts for any curve in the base $M$. Indeed, given a curve $\sigma(t)$, $t \in [0, 1]$, in $M$, there exists a unique horizontal lift $\tilde{\sigma}$ in $P$, given an initial point $u_0 \in P$, say $\tilde{\sigma}(0) = u_0$, where $\rho(u_0) = \sigma(0)$.

This results permits to construct a mapping

$$\rho^{-1}(x_0) \longrightarrow \rho^{-1}(x_1),$$

where $x_0 = \sigma(0)$ and $x_1 = \sigma(1)$, which will be called the *parallel transport along* $\sigma$. Notice that the parallel transport commutes with the action of the structure group.

### C.3.0.4 *Linear connections*

We have introduced the linear frame bundle $FM$ of a manifold $M$, which plays an essential role in this book. As we know, $FM$ is a principal bundle over $M$ with structure group $(Gl(n, \mathbb{R})$, where $n = \dim M)$.

**Definition C.3.3.** A linear connection on $M$ is a connection in the linear frame bundle $FM$.

Therefore, given a linear connection $\Gamma$ on $M$, we have a connection form

$$\omega : T(FM) \longrightarrow \mathfrak{gl}(n, \mathbb{R})$$

with curvature form

$$\Omega = d\omega \circ h.$$

### C.3.0.5 *Torsion*

Let $FM$ be the linear frame bundle of $M$, where $\dim M = n$.

We define the canonical form $\theta$ on $FM$ which is the $\mathbb{R}^n$-valued 1-form given by

$$\theta(X) = u^{-1}(T\rho_M(X)),$$

for $X \in T_u(FM)$, and the linear frame $u$ at a point $x \in M$ is understood as a linear isomorphism $u : \mathbb{R}^n \longrightarrow T_x M$.

The torsion form $\Theta$ of a linear connection $\Gamma$ is defined by

$$\Theta = d\theta \circ h.$$

If $\omega$ is the connection form, we can prove the following structure equations:

$$d\theta = -\omega \wedge \theta + \Theta,$$
$$d\omega = -\omega \wedge \omega + \Omega,$$

where we are using the Lie bracket in the Lie algebra of the general linear group $Gl(n, \mathbb{R})$.

### C.3.0.6   *Associated bundles*

Let $P(M, G)$ be a principal bundle and $F$ a manifold on which $G$ acts on the left. We will construct a fibre bundle $E$ over $M$ with standard fibre $F$ as follows:

- $G$ acts on $P \times F$:

$$g(u, \xi) = (ug, g^{-1}\xi);$$

- $E$ is just the quotient manifold

$$E = (P \times F)/G;$$

- the canonical projection $\pi : E \longrightarrow M$ is given by

$$\pi(e) = \pi([u, \xi]) = \rho(u).$$

$E$ will be called a fibre bundle associated with $P$.

**Example C.3.4 (The tangent bundle).** The tangent bundle $TM$ of $M$ is an associated vector bundle to the linear frame bundle $FM$ with standard fibre $\mathbb{R}^n$ and the usual action of matrices on vectors, where $n = \dim M$.

Indeed, $Gl(n, \mathbb{R})$ acts on $\mathbb{R}^n$ by the left in the canonical manner. So, any element $v \in (FM \times \mathbb{R}^n)/Gl(n, \mathbb{R})$ is a class of equivalence $v = [(u, \xi)]$, where $u$ is a linear reference at a point $x \in M$ and $\xi \in \mathbb{R}^n$. So, considering the image $u(\xi)$ we may prove that, in fact, this vector bundle is isomorphic to the tangent bundle $TM$.

If $E = (P \times F)/G$ is an associated bundle to $P(M, G)$ with standard fibre $F$, we can construct diffeomorphisms between $F$ and any fibre $E_x$ over $x$. Indeed, if $u \in P$ $\rho(u) = x$, then the mapping

$$\bar{u} : F \longrightarrow E_x,$$

$$\xi \mapsto \bar{u}(\xi) = [(u, \xi)]$$

is a diffeomorphism.

In the particular case of the tangent bundle $TM$, $\bar{u}$ is just the construction of the tangent vector $\bar{u}(\xi) \in T_x M$ whose components with respect to the basis $u$ are just $\xi = (\xi^1, \ldots, \xi^n) \in \mathbb{R}^n$.

### C.3.0.7   *Covariant derivatives associated to connections in principal bundles*

Assume that $\Gamma$ is a connection in the principal bundle $P(M, G)$, and $\pi : E \longrightarrow M$ is a vector bundle with standard fibre $F$ associated to $P$.

First of all, we can construct a parallel transport among the fibres of $E$ as follows.

Let $\sigma(t)$ a curve in $M$, $t \in [a, b]$, and $\tilde{\sigma}(t)$ its horizontal lift to $P$ (with respect to $\Gamma$). Then, for a fixed $\xi \in F$, $\bar{\sigma}(t) = [(\tilde{\sigma}(t), \xi)]$ is the horizontal lift of $\sigma(t)$ to $E$.

Now, assume that $\phi : M \longrightarrow E$ is a section of $\pi : E \longrightarrow M$ defined along $\sigma(t)$, that is, $\pi(\phi(\sigma(t))) = \sigma(t)$. Then we define the covariant derivative $\nabla_{\dot{\sigma}(t)} \phi$ of $\phi$ in the direction of $\dot{\sigma}(t)$ by

$$\nabla_{\dot{\sigma}(t)} \phi = \lim_{h \to 0} \frac{1}{h} \left[ \tau_t^{t+h}(\phi(\sigma(t+h))) - \phi(\sigma(t)) \right],$$

where

$$\tau_t^{t+h} : E_{\sigma(t+h)} = \pi^{-1}(\sigma(t+h)) \longrightarrow E_{\sigma(t)} = \pi^{-1}(\sigma(t))$$

is the parallel transport between these fibres along $\sigma$ from $\sigma(t+h)$ to $\sigma(t)$.

So,

$$\nabla_{\dot{\sigma}(t)} \phi \in E_{\sigma(t)} = \pi^{-1}(\sigma(t))$$

is another section of $E$ along $\sigma(t)$.

Let $X \in T_x M$ be a tangent vector. We define $\nabla_X \phi$ for a section $\phi$ by the formula

$$\nabla_X \phi = \nabla_{\dot{\sigma}(0)} \phi,$$

where $\sigma(t)$, $-\epsilon < t < \epsilon$ is a curve in $M$ such that $\sigma(0) = x$ and $\dot{\sigma}(0) = X$.

The extension of the expression $\nabla_X \phi$ where $X$ is a vector field on $M$ is natural:

$$(\nabla_X \phi)(x) = \nabla_{X(x)} \phi$$

for any $x \in M$.

Assume now that $P = FM$, $\Gamma$ is a linear connection and $TM$ its associated tangent bundle. Then, the above construction

$$\nabla_X Y$$

for any two vector fields $X$ and $Y$ on $M$, is just a covariant derivative on $M$.

The reader can check that the opposite construction also works: given a covariant derivative $\nabla$ on $M$, we can construct a linear connection whose associated covariant derivative is just $\nabla$. So, we recover the definition of linear connection given in Appendix B (Definition B.0.2).

**Remark C.3.5.** The above construction can be extended for any associated vector bundle to any arbitrary principal bundle $P(M, G)$.

### C.3.0.8  *Torsion and curvature*

Given a linear connection $\Gamma$ in $FM$, we have the following differential forms:

- $\omega$: connection form with values in $\mathfrak{gl}(n, \mathbb{R})$;
- $\Omega$: curvature form with values in $\mathfrak{gl}(n, \mathbb{R})$;
- $\theta$: canonical form with values in $\mathbb{R}^n$;
- $\Theta$: torsion form with values in $\mathbb{R}^n$.

The torsion and curvature forms are directly related to the tensor and curvature tensors of the associated covariant derivative $\nabla$ to $\Gamma$:

- $T(X, Y) = u(2\Theta(X^H, Y^H))$,
- $R(X, Y)Z = u(2\Omega(X^H, Y^H)(u^{-1}Z))$,

for all $X, Y, Z \in T_x M, u \in FM, \rho(u) = x$.

### C.3.0.9  *Local expressions*

Assume that $e_1, \ldots, e_n$ is the canonical basis of $\mathbb{R}^n$, and $E^i_j$ the canonical basis of $\mathfrak{gl}(n, \mathbb{R})$. Then

$$\theta = \theta^i e_i$$

and, if $\omega$ is the connection form of a linear connection $\Gamma$ on $M$, then

$$\omega = \omega^i_j E^j_i,$$
$$\Theta = \Theta^i e_i,$$
$$\Omega = \Omega^i_j E^j_i,$$

where $\theta^i, \omega^i_j, \Theta^i$ and $\Omega^i_j$ are real differential forms.

Moreover, if we take bundle coordinates in $FM$, $(x^i, x^i_j)$, we obtain, after a straightforward computation, the following formula:

$$\theta^i = y^i_j dx^j$$

where $(y^i_j)$ is the inverse matrix of $(x^i_j)$.

Consider now the section $\sigma$ defined on a coordinate neighborhood $(U, x^i)$ by

$$\sigma(x^i) = (x^i, \delta^i_j(x)),$$

that is,

$$\sigma(x) = \left( \frac{\partial}{\partial x^1}_{|x}, \ldots, \frac{\partial}{\partial x^n}_{|x} \right).$$

then, if we denote $\omega_U = \sigma^*\omega$, we have

$$\omega_U = (\Gamma^i_{jk} dx^j) E^k_i,$$

where $\Gamma^i_{jk}$ are just the Christoffel symbols of the covariant derivative corresponding to $\omega$.

# Bibliography

Abdelghani, Z. (2000). On gromov's theory of rigid transformation groups: a dual approach, *Ergodic Theory Dynam. Syst.* **20**, 3, pp. 935–946.

Almeida, R. and Kumpera, A. (1981). Structure produit dans la catégorie des algébroides de lie, *An. Acad. Brasil. Ciênc.* **53**, pp. 247–250.

Almeida, R. and Molino, P. (1985). Suites d'Atiyah et feuilletages transversalement complets, *C. R. Acad. Sci. Paris Sér. I Math.* **300**, 1, pp. 13–15.

B. A. Bilby (1960). Continuous distributions of dislocations, in *Progress in Solid Mechanics*, Vol. 1 (North-Holland Publishing Co., Amsterdam), pp. 329–398.

Bernard, D. (1960). Sur la géométrie différentielle des $G$-structures, *Ann. Inst. Fourier Grenoble* **10**, pp. 151–270, http://www.numdam.org/item?id=AI F_1960_10_151_0.

Bloom, F. (1979). *Modern Differential Geometric Techniques in the Theory of Continuous Distributions of dislocations*, Lecture Notes in Mathematics, Vol. 733 (Springer, Berlin).

Brandt, H. (1927). Über eine Verallgemeinerung des Gruppenbegriffes, *Math. Ann.* **96**, 1, pp. 360–366, https://doi.org/10.1007/BF01209171.

Brown, R. (1987). From groups to groupoids: a brief survey, *Bull. London Math. Soc.* **19**, 2, pp. 113–134, doi:10.1112/blms/19.2.113, http://oup.prod.sis.lan/blms/article-pdf/19/2/113/772134/19-2-113.pdf, h ttps://doi.org/10.1112/blms/19.2.113.

Cámpos, C. M., Epstein, M., and de León, M. (2008). Functionally graded madia, *Int. J. Geom. Methods Mod. Phys.* **05**, 03, pp. 431–455, doi:10.1142/S0219887808002874, https://doi.org/10.1142/S0219887808002 874, https://doi.org/10.1142/S0219887808002874.

Chern, S. S. (1966). The geometry of $G$-structures, *Bull. Amer. Math. Soc.* **72**, pp. 167–219, https://doi.org/10.1090/S0002-9904-1966-11473-8.

Ciaglia, F. M., A. Ibort, A., and Marmo, G. (2018). A gentle introduction to Schwinger's formulation of quantum mechanics: the groupoid picture, *Modern Phys. Lett. A* **33**, 20, pp. 1850122, 8, doi:10.1142/S0217732318501225, https://doi.org/10.1142/S0217732318501225.

Coleman, B. D. (1965). Simple liquid crystals, *Arch. Rational Mech. Anal.* **20**, pp. 41–58, doi:10.1007/BF00250189.

Cordero, L. A., Dodson, C. T. J., and de León, M. (1989). *Differential Geometry of Frame Bundles*, Mathematics and its Applications, Vol. 47 (Kluwer Academic Publishers Group, Dordrecht), https://doi.org/10.1007/978-94-009-1265-6.

Cortés, J., de León, M., Marrero, J. C., de Diego, D. M., and Martínez, E. (2006). A survey of Lagrangian mechanics and control on Lie algebroids and groupoids, *Int. J. Geom. Methods Mod. Phys.* **3**, 3, pp. 509–558, doi:10.1142/S0219887806001211, https://doi.org/10.1142/S0219887806001211.

Crainic, M. and Fernandes, R. L. (2003). Integrability of Lie brackets, *Ann. of Math.* **157**, 2, pp. 575–620, https://doi.org/10.4007/annals.2003.157.575.

de Diego, D. M. and Sato Martín de Almagro, R. (2018). Variational order for forced Lagrangian systems, *Nonlinearity* **31**, 8, pp. 3814–3846, doi:10.1088/1361-6544/aac5a6, https://doi.org/10.1088/1361-6544/aac5a6.

de León, M., Marrero, J. C., and de Diego, D. M. (2010). Linear almost poisson structures and Hamilton-Jacobi equation. applications to nonholonomic mechanics, *J. Geom. Mech.* **2**, 2, pp. 159–198.

de León, M. and Rodrigues, P. R. (1989). *Methods of Differential Geometry in Analytical Mechanics*, North-Holland Mathematics Studies (Elsevier Science, Amsterdam), https://books.google.es/books?id=5pCfP8CiSzAC.

Dufour, J. P. and Zung, N. T. (2005). *Poisson Structures and Their Normal Forms*, Progress in Mathematics, Vol. 242 (Birkhäuser Verlag, Basel).

Duistermaat, J. J. and Kolk, J. A. C. (2000). *Lie Groups*, Universitext (Springer-Verlag, Berlin), https://doi.org/10.1007/978-3-642-56936-4.

Ehresmann, C. (1952). Les prolongements d'une variété différentiable. V. Covariants différentiels et prolongements d'une structure infinitésimale, *C. R. Acad. Sci. Paris* **234**, pp. 1424–1425.

Ehresmann, C. (1953). Introduction à la théorie des structures infinitésimales et des pseudogroupes de Lie, in *Colloque de topologie et géométrie différentielle, Strasbourg, 1952, No. 11* (La Bibliothèque Nationale et Universitaire de Strasbourg), p. 16.

Ehresmann, C. (1954). Extension du calcul des jets aux jets non holonomes, *C. R. Acad. Sci. Paris* **239**, pp. 1762–1764.

Ehresmann, C. (1955a). Applications de la notion de jet non holonome, *C. R. Acad. Sci. Paris* **240**, pp. 397–399.

Ehresmann, C. (1955b). Les prolongements d'un espace fibré différentiable, *C. R. Acad. Sci. Paris* **240**, pp. 1755–1757.

Ehresmann, C. (1956). sur les connexions d'ordre supérieur, in *Dagli Atti del V Congresso dell'Unione Matematica Italiana*, pp. 344–346.

Ehresmann, C. (1959). Catégories topologiques et catégories différentiables, in *Colloque Géom. Diff. Globale (Bruxelles, 1958)* (Centre Belge Rech. Math., Louvain), pp. 137–150.

Ehresmann, C. (1995). Les connexions infinitésimales dans un espace fibré différentiable, in *Séminaire Bourbaki, Vol. 1* (Soc. Math. France, Paris), Exp. No. 24, pp. 153–168.

Eilenberg, S. and MacLane, S. (1945). General theory of natural equivalences, *Trans. Amer. Math. Soc.* **58**, pp. 231–294, https://doi.org/10.2307/1990 284.

Elżanowski, M., Epstein, M., and Śniatycki, J. (1990). *G*-structures and material homogeneity, *J. Elasticity* **23**, 2–3, pp. 167–180, https://doi.org/10.1007/BF00054801.

Elżanowski, M. and Prishepionok, S. (1992). Locally homogeneous configurations of uniform elastic bodies, *Rep. Math. Phys.* **31**, 3, pp. 329–340, https://doi.org/10.1016/0034-4877(92)90023-T.

Epstein, M. (2010). *The Geometrical Language of Continuum Mechanics* (Cambridge University Press, Cambridge), doi:10.1017/CBO9780511762673.

Epstein, M. (2017). Laminated uniformity and homogeneity, *Mechanics Research Communications* doi:https://doi.org/10.1016/j.mechrescom.2017.05.004, http://www.sciencedirect.com/science/article/pii/S0093641317302136.

Epstein, M. and de León, M. (1998). Geometrical theory of uniform cosserat media, *J. Geom. Phys.* **26**, 1–2, pp. 127–170, https://doi.org/10.1016/S03 93-0440(97)00042-9.

Epstein, M. and de León, M. (2000). Homogeneity without uniformity: towards a mathematical theory of functionally graded materials, *International Journal of Solids and Structures* **37**, 51, pp. 7577–7591, doi:https://doi.org/10.1016/S0020-7683(99)00309-1, http://www.sciencedi rect.com/science/article/pii/S0020768399003091.

Epstein, M. and de León, M. (2016). Unified geometric formulation of material uniformity and evolution, *Math. Mech. Complex Syst.* **4**, 1, pp. 17–29, https://doi.org/10.2140/memocs.2016.4.17.

Epstein, M. and Elzanowski, M. (2007). *Material Inhomogeneities and their Evolution: A Geometric Approach*, Interaction of Mechanics and Mathematics (Springer, Berlin).

Epstein, M., Jiménez, V. M., and de León, M. (2019). Material geometry, *Journal of Elasticity* **135**, 1, pp. 237–260, doi:10.1007/s10659-018-9693-2, https://doi.org/10.1007/s10659-018-9693-2.

Eringen, A. C. (1972a). Nonlocal polar elastic continua, *Int. J. Eng. Sci.* **10**, 1, pp. 1–16, doi:https://doi.org/10.1016/0020-7225(72)90070-5, http://www.sciencedirect.com/science/article/pii/0020722572900705.

Eringen, A. C. (1972b). Theory of micromorphic materials with memory, *Int. J. Eng. Sci.* **10**, 7, pp. 623–641, doi:https://doi.org/10.1016/0020-7225(72)90089-4, http://www.sciencedirect.com/science/article/pii/00207 22572900894.

Eringen, A. C. (1983). On differential equations of nonlocal elasticity and solutions of screw dislocation and surface waves, *J. Appl. Phys.* **54**, 9, pp. 4703–4710,

doi:10.1063/1.332803, https://doi.org/10.1063/1.332803, https://doi.org/1
0.1063/1.332803.

Eringen, A. C. (2002). *Nonlocal Continuum Field Theories* (Springer-Verlag,
New York).

Eshelby, J. D. (1951). The force on an elastic singularity, *Philos. Trans. Royal
Soc. London A: Math. Phys. Eng. Sci.* **244**, 877, pp. 87–112, doi:10.1098/
rsta.1951.0016, http://rsta.royalsocietypublishing.org/content/244/877/87
.full.pdf, http://rsta.royalsocietypublishing.org/content/244/877/87.

Ferraro, S., de León, M., Marrero, J., de Diego, D. M., and Vaquero, M. (2017).
On the geometry of the Hamilton–Jacobi equation and generating functions,
*Arch. Rational Mech. Anal.* **226**, pp. 226–243.

Fujimoto, A. (1972). *Theory of G-structures* (Study Group of Geometry, Department of Applied Mathematics, College of Liberal Arts and Science,
Okayama University, Okayama), English Edition, translated from the original Japanese, Publications of the Study Group of Geometry, Vol. 1.

Higgins, P. J. and Mackenzie, K. C. H. (1990). Algebraic constructions in the
category of Lie algebroids, *J. Algebra* **129**, 1, pp. 194–230, https://doi.org
/10.1016/0021-8693(90)90246-K.

Hirsch, M. W. (1994). *Differential topology, Graduate Texts in Mathematics*, Vol. 33 (Springer-Verlag, New York), corrected reprint of the 1976
original.

Jiménez, V. M., de León, M., and Epstein, M. (2017). Material distributions, *Mathematics and Mechanics of Solids* **25**, 7, pp. 1450–1458,
doi:10.1177/1081286517736922, https://doi.org/10.1177/108128651773692
2, https://doi.org/10.1177/1081286517736922.

Jiménez, V. M., de León, M., and Epstein, M. (2018). Characteristic distribution: An application to material bodies, *J. Geom. Phys.* **127**, pp. 19–31,
doi:https://doi.org/10.1016/j.geomphys.2018.01.021, http://www.scienced
irect.com/science/article/pii/S0393044018300378.

Jiménez, V. M., de León, M., and Epstein, M. (2018). Lie groupoids and
algebroids applied to the study of uniformity and homogeneity of
cosserat media, *Int. J. Geom. Methods Mod. Phys.* **15**, 08, p. 1830003,
doi:10.1142/S0219887818300039, https://doi.org/10.1142/S0219887818300
039, https://doi.org/10.1142/S0219887818300039.

Jiménez, V. M., de León, M., and Epstein, M. (2019). Lie groupoids and algebroids
applied to the study of uniformity and homogeneity of material bodies,
*J. Geom. Mech.* **11**, 3, pp. 301–324, doi:10.3934/jgm.2019017.

Jiménez, V. M., de León, M., and Epstein, M. (2020). *On the Homogeneity of
Non-uniform Material Bodies* (Springer International Publishing, Cham),
pp. 381–416, doi:10.1007/978-3-030-42683-5_9, https://doi.org/10.1007/97
8-3-030-42683-5_9.

Johnson, W. W. and Story, W. E. (1879). Notes on the "15" puzzle, *Amer. J.
Math.* **2**, 4, pp. 397–404, http://www.jstor.org/stable/2369492.

Kobayashi, S. I. and Nomizu, K. (1996). *Foundations of Differential Geometry.*
Vol. I, Wiley Classics Library (John Wiley & Sons, Inc., New York), reprint
of the 1963 original, A Wiley-Interscience Publication.

Kondo, K. (1955). *Geometry of Elastic Deformation and Incompatibility, in Memoirs of the Unifying Study of the Basic Problems in Engineering Science by Means of Geometry*, edited by Kondo, K. (Division C, Gakujutsu Bunken Fukyo-Kai **1**, pp. 5–17.

Kosmann-Schwarzbach, Y. and Mackenzie, K. C. H. (2002). Differential operators and actions of Lie algebroids, in *Quantization, Poisson Brackets and Beyond (Manchester, 2001)*, Contemporary Mathematics, Vol. 315 (American Mathematical Society, Providence, RI), pp. 213–233, https://doi.org/10.1090/conm/315/05482.

Kriegl, A. and Michor, P. W. (1997). *The convenient setting of global analysis, Mathematical Surveys and Monographs*, Vol. 53 (American Mathematical Society, Providence, RI), doi:10.1090/surv/053, https://doi.org/10.1090/surv/053.

Kröner, E. (1960). *Allgemeine Kontinuumstheorie der Versetzungen und Eigenspannungen*, Vol. 4, https://books.google.es/books?id=bXCdGwAACAAJ.

Kröner, E. (1968). *Mechanics of Generalized Continua* (Springer, Heidelberg).

Lardner, R. W. (1974). *Mathematical Theory of Dislocations and Fracture*, Mathematical Expositions (University of Toronto Press, Toronto), https://books.google.es/books?id=WZlsAAAAMAAJ.

Lee, J. M. (2000). *Introduction to Topological Manifolds, Graduate Texts in Mathematics* (Springer), https://books.google.es/books?id=5LqQgkS3--MC.

Mackenzie, K. C. H. (1987). *Lie Groupoids and Lie Algebroids in Differential Geometry*, London Mathematical Society Lecture Note Series, Vol. 124 (Cambridge University Press, Cambridge), https://doi.org/10.1017/CBO9780511661839.

Mackenzie, K. C. H. (2005). *General Theory of Lie Groupoids and Lie Algebroids*, London Mathematical Society Lecture Note Series, Vol. 213 (Cambridge University Press, Cambridge), 0-521-49928-3, https://doi.org/10.1017/CBO9781107325883.

Mackenzie, K. C. H. and Xu, P. (2000). Integration of Lie bialgebroids, *Topology* **39**, 3, pp. 445–467, https://doi.org/10.1016/S0040-9383(98)00069-X.

Marrero, J. C., de Diego, D. M., and Martínez, E. (2006). Discrete lagrangian and hamiltonian mechanics on lie groupoids, *Nonlinearity* **19** (2006), no. 6, 1313–1348, doi: https://doi.org/10.1088/0951-7715/19/6/006.

Marsden, J. E. and Hughes, T. J. R. (1994). *Mathematical Foundations of Elasticity* (Dover Publications, Inc., New York), corrected reprint of the 1983 original.

Maugin, G. A. (1993). *Material Inhomogeneities in Elasticity, Applied Mathematics and Mathematical Computation*, Vol. 3 (Chapman & Hall, London), https://doi.org/10.1007/978-1-4899-4481-8.

Michor, P. W. (2008). *Topics in Differential Geometry, Graduate Studies in Mathematics*, Vol. 93 (American Mathematical Society, Providence), https://doi.org/10.1090/gsm/093.

Moerdijk, I. and Mrčun, J. (2002). On integrability of infinitesimal actions, *Amer. J. Math.* **124**, 3, pp. 567–593, http://muse.jhu.edu/journals/american_journal_of_mathematics/v124/124.3moerdijk.pdf.

Moerdijk, I. and Mrčun, J. (2006). On the integrability of Lie subalgebroids, *Adv. Math.* **204**, 1, pp. 101–115, https://doi.org/10.1016/j.aim.2005.05.011.

Nabarro, F. R. N. (1987). *Theory of Crystal Dislocations*, Dover Books on Physics and Chemistry (Dover Publications, New York), https://books.google.es/books?id=zD5CAQAAIAAJ.

Noll, W. (1954). *On the Continuity of the Solid and Fluid States* (ProQuest LLC, Ann Arbor, MI), http://gateway.proquest.com/openurl?url_ver=Z39.88-2004&rft_val_fmt=info:ofi/fmt:kev:mtx:dissertation&res_dat=xri:pqdiss&rft_dat=xri:pqdiss:0010155, Ph.D., thesis Indiana University.

Noll, W. (1967). Materially uniform simple bodies with inhomogeneities, *Arch. Rational Mech. Anal.* **27**, pp. 1–32, https://doi.org/10.1007/BF00276433.

O'Neill, B. (1983). *Semi-Riemannian Geometry With Applications to Relativity*, Pure and Applied Mathematics (Elsevier Science).

Pradines, J. (1966). Théorie de Lie pour les groupoïdes différentiables. Relations entre propriétés locales et globales, *C. R. Acad. Sci. Paris Sér. A–B* **263**, pp. A907–A910.

Pradines, J. (1967). Théorie de Lie pour les groupoïdes différentiables. Calcul différenetiel dans la catégorie des groupoïdes infinitésimaux, *C. R. Acad. Sci. Paris Sér. A–B* **264**, pp. A245–A248.

Pradines, J. (1968a). Géometrie différentialle au-dessus d'un groupoïde, *C. R. Acad. Sci. Paris Sér. A* **266**, pp. 1194–1196.

Pradines, J. (1968b). Troisième théorème de Lie pour les groupoïdes différentiables, *C. R. Acad. Sci. Paris Sér. A–B* **267**, pp. A21–A23.

Ramsay, A. and Renault, J. (eds.) (2001). *Groupoids in analysis, geometry, and physics*, Contemporary Mathematics, Vol. 282 (American Mathematical Society, Providence, RI), doi:10.1090/conm/282, https://doi.org/10.1090/conm/282, papers from the AMS-IMS-SIAM Joint Summer Research Conference held at the University of Colorado, Boulder, CO, June 20–24, 1999.

Reinhart, B. L. (1983). *Differential Geometry of Foliations: The Fundamental Integrability Problem*, Ergebnisse der Mathematik und ihrer Grenzgebiete (Springer-Verlag), https://books.google.es/books?id=LMMPAQAAMAAJ.

Segev, R. (1986). Forces and the existence of stresses in invariant continuum mechanics, *J. Math. Phys.* **27**, pp. 163–170.

Stefan, P. (1974). Accessible sets, orbits, and foliations with singularities, *Proc. London Math. Soc.* (3) **29**, pp. 699–713, https://doi.org/10.1112/plms/s3-29.4.699.

Sternberg, S. (1983). *Lectures on differential geometry*, 2nd edn. (Chelsea Publishing Co., New York), with an appendix by Sternberg and Victor W. Guillemin.

Sussmann, H. J. (1973). Orbits of families of vector fields and integrability of distributions, *Trans. Amer. Math. Soc.* **180**, pp. 171–188, https://doi.org/10.2307/1996660.

Valdés, J. N., Villalón, Á. F. T., and Alarcón, J. A. V. (2006). *Elementos de la teoría de grupoides y algebroides* (Universidad de Cádiz, Servicio de Publicaciones, Cádiz), https://books.google.es/books?id=srgOQIoHqfMC.

Volterra, V. (1907). Sur l'équilibre des corps élastiques multiplement connexes, *Ann. Sci. École Norm. Sup.* (3) **24**, pp. 401–517, http://www.numdam.or g/item?id=ASENS_1907_3_24__401_0.

Wang, C. C. (1965). A general theory of subfluids, *Arch. Rational Mech. Anal.* **20**, pp. 1–40, doi:10.1007/BF00250188.

Wang, C. C. (1967). On the geometric structures of simple bodies. A mathematical foundation for the theory of continuous distributions of dislocations, *Arch. Rational Mech. Anal.* **27**, pp. 33–94, https://doi.org/10.1007/BF00276434.

Wang, C. C. and Truesdell, C. (1973). *Introduction to Rational Elasticity* (Noordhoff International Publishing, Leyden), monographs and Textbooks on Mechanics of Solids and Fluids: Mechanics of Continua.

Weinstein, A. (1996). Lagrangian mechanics and groupoids, in *Mechanics Day* (Waterloo, ON, 1992), Fields Inst. Commun., Vol. 7 (American Mathematical Society, Providence, RI), pp. 207–231.

Weinstein, A. (2001). Groupoids: unifying internal and external symmetry. A tour through some examples, in *Groupoids in Analysis, Geometry, and Physics* (Boulder, CO, 1999), Contemporary Mathematics, Vol. 282 (Amer. Math. Soc., Providence, RI), pp. 1–19, doi:10.1090/conm/282/04675, https://doi .org/10.1090/conm/282/04675.

Zakrzewski, S. (1990a). Quantum and classical pseudogroups. i. union pseudogroups and their quantization, *Comm. Math. Phys.* **134**, 2, pp. 347–370, https://projecteuclid.org:443/euclid.cmp/1104201734.

Zakrzewski, S. (1990b). Quantum and classical pseudogroups. ii. differential and symplectic pseudogroups, *Comm. Math. Phys.* **134**, 2, pp. 371–395, https: //projecteuclid.org:443/euclid.cmp/1104201735.

# Index

Printed in the United States
by Baker & Taylor Publisher Services